Engineering Materials and Processes

Series Editor

Professor Brian Derby, Professor of Materials Science
Manchester Materials Science Centre, Grosvenor Street, Manchester, M1 7HS, UK

Other titles published in this series

Fusion Bonding of Polymer Composites
C. Ageorges and L. Ye

Composite Materials
D.D.L. Chung

Titanium
G. Lütjering and J.C. Williams

Corrosion of Metals
H. Kaesche

Corrosion and Protection
E. Bardal

Intelligent Macromolecules for Smart Devices
L. Dai

Microstructure of Steels and Cast Irons
M. Durand-Charre

Phase Diagrams and Heterogeneous Equilibria
B. Predel, M. Hoch and M. Pool

Computational Mechanics of Composite Materials
M. Kamiński

Gallium Nitride Processing for Electronics, Sensors and Spintronics
S.J. Pearton, C.R. Abernathy and F. Ren

Materials for Information Technology
E. Zschech, C. Whelan and T. Mikolajick

Fuel Cell Technology
N. Sammes

Casting: An Analytical Approach
A. Reikher and M.R. Barkhudarov

Computational Quantum Mechanics for Materials Engineers
L. Vitos

Modelling of Powder Die Compaction
P.R. Brewin, O. Coube, P. Doremus and J.H. Tweed

Silver Metallization
D. Adams, T.L. Alford and J.W. Mayer

Microbiologically Influenced Corrosion
R. Javaherdashti

Modeling of Metal Forming and Machining Processes
P.M. Dixit and U.S. Dixit

Electromechanical Properties in Composites Based on Ferroelectrics
V.Yu. Topolov and C.R. Bowen

Charged Semiconductor Defects
E.G. Seebauer and M.C. Kratzer

Modelling Stochastic Fibrous Materials with Mathematica®
W.W. Sampson

Ferroelectrics in Microwave Devices, Circuits and Systems
S. Gevorgian

Markus Niederberger • Nicola Pinna

Metal Oxide Nanoparticles in Organic Solvents

Synthesis, Formation, Assembly and Application

Springer

Prof. Dr. Markus Niederberger
Swiss Federal Institute of Technology
(ETH) Zurich
Department of Materials
Laboratory for Multifunctional Materials
Wolfgang-Pauli-Strasse 10
8093 Zurich
Switzerland
markus.niederberger@mat.ethz.ch

Prof. Dr. Nicola Pinna
University of Aveiro
Department of Chemistry, CICECO
Campus Universitario de Santiago
3810-193 Aveiro
Portugal
pinna@ua.pt

and

Seoul National University (SNU)
College of Engineering
School of Chemical
and Biological Engineering
151-744 Seoul
Korea
pinna@snu.ac.kr

ISSN 1619-0181
ISBN 978-1-4471-2509-9
e-ISBN 978-1-84882-671-7
DOI 10.1007/978-1-84882-671-7
Springer Dordrecht Heidelberg London New York

British Library Cataloguing in Publication Data
A catalogue record for this book is available from the British Library

Softcover reprint of the hardcover 1st edition 2009

Cover design: eStudioCalamar, Figueres/Berlin

Printed on acid-free paper

Springer is part of Springer Science+Business Media (www.springer.com)

To our wives Anja and Sabina

Foreword

Some scientists, especially those far off materials sciences, still perceive the prefix "nano" as a hype. To my opinion, this notation might describe the vigorous expansion phase of nanosciences and nanotechnology, but the hype meanwhile turned into real science, into real products and jobs.

Contrary to previous hyped topics, the nano-community was able to fill many promises with reality, at least to a larger extent, and this is why nanosciences are still flourishing, still growing, partly even beyond expectations of the educated experts.

A key part of nanosciences is nanoparticulate matter, including nanoparticles, nanorods, nanoplates, or even more complex tectonic assemblies. Exactly on these examples we can nicely illustrate what nanosciences really provides. Simply by finely dispersing common bulk materials down to the nanoscale, new properties can be observed: inert materials become catalysts, insulators become conductors, white substances become transparent, stable materials become combustible. It is known since the ancient world: Gold nanoparticles are shining red or blue, and the beauty of medieval church windows is based on the extraordinary color strength and the high thermal stability of those dispersions, exceeding ordinary dyes by some orders of magnitude.

There is more than only plasmonic color to be discovered: There is a whole new world out there in the nano cosmos, a world which already Wolfgang Ostwald about a hundred years ago named the "world of lost dimension". Some colleagues even regard nanostructures as a new adjustable state of matter. This is due to the fact that atoms at surfaces behave differently to those in the bulk, and nanoparticles are literally dominated by those surfaces.

It is a save prediction: The new questions to face while analyzing alternative energy cycles of society, communication technology, the desire for new catalysts for more efficient chemical reactions, new light sources, or better performing construction materials will only be answered by implementation of nanostructural aspects. There will be no real alternative or choice.

For many of those applications, appropriate nanostructures however do not yet exist, or their production is not sustainable so that "de novo" systems and their synthetic pathways have to be designed from scratch. This is where the present book sets in: Both authors are most competent in this field, pioneers of a synthetic route towards metal oxide nanoparticles, which is meanwhile called "nonaqueous sol-gel route". This technique not only turned out to be simple, convenient, scalable, mass efficient and rather sustainable, but it also was proven to be extremely flexible with respect to size and composition so that – within just a few years – an impressive number of binary, ternary and quaternary nanocrystalline oxides in a variety of sizes and morphologies could be assessed. This explosive development is nicely summarized and structured in this book, and this is why it is a timely endeavor.

The book not only reviews comprehensively the primary literature and observations, it also gives an actual description of current trends in the field and the still ongoing improvements of the technology, say by incorporation of microwave synthesis. Special emphasis is also put on the "social behavior" of nanoparticles, i.e., their ability to self-organize towards most complicated and aesthetically very appealing superstructures. If I look on some of those images, the word "emergence" is getting a real meaning even in the otherwise "dead" inorganic world. Some exemplary model cases delineating the properties and applications of nanoparticles are completing the story.

I think, the book is an excellent compendium both for the expert reader (due to the comprehensive, systematic presentation of the data) as well as a reading for students and laymen to get "infected" with the real fascination and potential of the "nanos". Beside the wish for controlling and accessing the unknown, it is also the beauty of the found and the joy of the discovery which makes science so attractive. The following book is rich in that!

Potsdam, *Prof. Dr. Markus Antonietti*
February 2009

Preface

"Oxide Synthesis as Cornerstone of Nanoscience" – This statement was the title of a short Editorial we recently wrote for a special issue of the European Journal of Inorganic Chemistry on Metal Oxide Nanoparticles (Eur. J. Inorg. Chem. 2008, 825). Without any doubts, metal oxide nanoparticles play an outstanding role in many applications that are regarded as particularly promising within the broad area of Nanotechnology, e.g., nanophotonics, spintronics, energy storage and conversion, catalysis, or biomedical applications.

The great variety of structures and properties of metal oxides made this class of materials not only the primary target in solid state chemistry, but also represents a major inspiration for designing new materials on the nanoscale. It is therefore not surprising that a large number of synthesis methodologies have been reported for the size and shape-controlled synthesis of metal oxide nanostructures. One of the most versatile and fast developing approaches are nonaqueous or nonhydrolytic synthesis protocols, i.e., syntheses performed in organic solvents under exclusion of water. The main challenge we faced during writing this book was that we wanted to present both general and basic principles of metal oxide nanoparticle research as well as a rather exhaustive overview of the various metal oxide nanoparticles synthesized in organic solvents so far. We solved the problem in such a way that the main text explains the concepts on selected examples, whereas several tables list the various metal oxides synthesized via nonaqueous processes. We put great efforts in the preparation of these tables, i.e., we included all the literature we were aware of (until the end of 2008), offering a unique information source for chemists, physicists, materials scientists, and engineers to find the appropriate synthesis method for a targeted metal oxide with the desired properties.

The contents of the various chapters in this book were chosen based on a personal priorization of the most fascinating topics in this research area. After a short and general excursion into the world of nanoparticles in Chapter 1, we discuss the basic principles of nonaqueous sol-gel chemistry in comparison with aqueous systems. Although water-based processes are generally preferred, in the case of metal oxide nanoparticle synthesis the use of organic solvents represents an advantageous alternative, which is elaborated in more detail in Chapter 2. Chapter 3 and 4 are fully dedicated to the synthesis of metal oxide nanoparticles and metal oxide-based organic-inorganic hybrids, involving surfactant-assisted (Chapter 3) and surfactant-free routes (Chapter 4). Chapter 5 presents the main chemical pathways leading to metal oxides in organic solvents. There is no doubt that in addition to the extensive synthesis work, also the assembly and positioning of nanoparticles in desired locations and across extended length scales as well as the in-depth investigation of the physical and chemical properties are key steps on the way to implement these materials into technological devices. We took these topics into account in Chapter 6, dealing with the assembly, Chapter 7 on the characterization and Chapter 8 on the properties and applications of metal oxide nanoparticles. The last Chapter of the book summarizes the current knowledge and the future challenges in the field of metal oxide nanoparticles prepared in organic solvents.

We are thankful to our current and past group members for their great scientific work and in particular to Guylhaine Clavel for proofreading the manuscript.

Zurich and Aveiro,
February 2009

Prof. Dr. Markus Niederberger
Prof. Dr. Nicola Pinna

Contents

Chapter 1
Introduction

1.1 Fascination and Motivation of Nanoparticle Research

Research on nanoparticles, including synthesis, characterization of the structural, chemical and physical properties, assembly into 1-, 2- and 3-dimensional architectures extending over several lengths scales and with hierarchical construction principles, and application in various fields of technology, represents a fundamental cornerstone of nanoscience and nanotechnology. Many different synthesis techniques gave access to nanomaterials with a wide range of compositions, well-defined and uniform crystallite sizes, extraordinary and unprecedented crystallite shapes, and complex assembly properties. Although gas-phase processes are successfully employed for the low-cost production of large quantities of nanopowders [7, 12, 25], it seems that liquid-phase syntheses are more flexible with regard to the controlled variation of structural, compositional, and morphological features of the final nanomaterials. Liquid-phase routes include coprecipitation, hydrolytic as well as nonhydrolytic sol-gel processes, hydrothermal or solvothermal methods, template synthesis and biomimetic approaches [3]. However, often the synthesis protocol for a targeted material involves not just one, but a combination of several of these methods.

The synthesis of particles with control over size, shape and size distribution is not a special feature of nanoscience, but has been an integral part of colloid chemistry for decades. However, for quite a while, most efforts invested in the preparation of uniform, finely dispersed particles were essentially considered as due to the curiosity of colloid scientists, and, with the exception of polymer colloids (latexes), without any practical significance [14, 15]. With the advent of nanoscience this attitude has changed completely. Nowadays, there is no doubt about the fundamental role of uniform powders in many areas of science and technology. Parallel to the development of highly advanced analytical tools, enabling the characterization of small structures with atomic resolution, the size of the targeted objects and devices decreased rapidly below the 100 nm limit. The preparation of nanostructures on such a small size scale makes great demands to the synthesis methodology and therefore it is a great challenge to develop a "synthetic chemistry" of nanoparticles that is as precise as that used to make molecules [27]. Two strategies are generally pursued to prepare nanostructures: the "top-down" approach, using physical methods such as photolithography and related techniques, and the "bottom-up" employing physical and chemical approaches. Miniaturization by lithographic techniques is mainly driven by microelectronics industry and has reached a level, at which feature sizes of around 10 nm can be generated [6, 26]. However, upon down-scaling the costs associated with lithography equipment and operating facilities rapidly increase, and the features achievable are on the one hand rather coarse and irregular on an atomic scale, and on the other hand difficult to extend onto non-planar surfaces or to 3D structures. The bottom-up approach, making use of both specific and non-covalent interactions (e.g. hydrogen bonding, electrostatic and van der Waals interactions) between molecules or colloidal particles to assemble discrete nanoscale structures, represents a valuable alternative on the way to further miniaturization of electronic devices and to the fabrication of complex 3D architectures [18, 28]. The use of nanoparticles as building blocks has the advantage that highly advanced synthesis methodologies provide control over crystallite size and shape with a precision well beyond that of top-down lithography. The bottom-up assembly of nanoparticles may prove to be a solution to the technological challenges faced by the semiconductor industry [13]. From a scientific point of view, the use of nanoparticles as "artificial atoms" to form 1-, 2- and 3-dimensional arrays extending over several length scales is particularly fascinating, because these superlattices are expected to have unprecedented and striking collective properties resulting from the interactions between the nanobuilding blocks that cannot be found in the individual constituents, thus opening up an almost indefinite playground to design novel multi-component materials [16, 22, 24, 29].

The availability of reproducible colloidal synthesis protocols for nanoparticles whose compositional, structural, morphological and surface characteristics are well-defined and uniform, is a prerequisite on the way to such complex superstructures. Research on the preparation of inorganic nanoparticles and nanostructures has always been, and still will be, at the heart of nanoscience for the next few years. The significance of nanoparticle synthesis is also reflected in the fact that many groundbreaking findings that can be regarded as milestones in the history of nanoscience are directly related to synthetic work, for example the discovery of carbon nanotubes [10], the synthesis of well-defined quantum dots [17], or the shape control of CdSe nanocrystals [19].

The size- and shape-dependent physical and chemical properties of semiconductor nanoparticles [9, 1, 2] as well as the increased surface-to-volume ratio of nanoscale materials in general raised expectations for a better performance of nanomaterials compared to their bulk counterparts in many applications. Intensive works on semiconductor-doped glasses [5], on the photocatalytic properties of colloidal CdS [11, 8] and on the "absorption of light in a semiconductor sphere" [4] finally resulted in the description of the quantum size effect more than 25 years ago. Brus et al. found that CdS crystallites in the size range of a few nanometers did not have the electronic spectra of the bulk material, even though they exhibited the same unit cell and bond length as the bulk material [23]. These findings opened up a new and exciting possibility to tailor the chemical and physical properties of a material: New applications and properties are a result of controlling crystallite size and shape on a nanometer scale rather than of altering the composition [2].

1.2 Metal Oxides: Properties and Applications

Among all the functional materials to be synthesized on the nanoscale, metal oxides are particularly attractive candidates, from a scientific as well as from a technological point of view. The unique characteristics of metal oxides make them the most diverse class of materials, with properties covering almost all aspects of materials science and solid state physics. The great variety of structures and properties made them the primary target in solid state chemistry and still gives inspiration for designing new materials. The crystal structures range from simple rock salt to highly complex incommensurately modulated structures, and the nature of the metal-oxygen bonding varies from nearly ionic to covalent or metallic [20, 21]. Associated with such changes in structure and bonding, oxidic materials exhibit fascinating electronic and magnetic properties. Some oxides like RuO_2 or ReO_3 are metallic, whereas $BaTiO_3$ for example is an insulator. The magnetic properties found in oxides include ferro-, ferri- or antiferromagnetic behavior. Some oxides possess switchable orientation states as in ferroelectrics (e.g., titanates, niobates, or tantalates). Other fascinating classes of materials within the metal oxide family are the cuprate superconductors, the manganites showing colossal magnetoresistance, or multiferroics combining ferroelectricity and ferromagnetism within the same material (e.g., $BiFeO_3$, $BiMnO_3$). This exceptionally broad spectrum of properties makes (bulk) metal oxides a vital constituent in technological applications like gas sensing, medical science, electronics, ceramics, energy conversion and storage, and surface coatings, just to mention a few (Figure 1.1).

Fig. 1.1. Selected applications of metal oxides

References

1. Alivisatos, A.P.: Perspectives on the physical chemistry of semiconductor nanocrystals. J. Phys. Chem. **100**, 13,226–13,239 (1996)
2. Burda, C., Chen, X., Narayanan, R., El-Sayed, M.A.: Chemistry and properties of nanocrystals of different shapes. Chem. Rev. **105**, 1025–1102 (2005)
3. Cushing, B.L., Kolesnichenko, V.L., O'Connor, C.J.: Recent advances in the liquid-phase syntheses of inorganic nanoparticles. Chem. Rev. **104**, 3893–3946 (2004)
4. Efros, A.L., Efros, A.L.: Interband absorption of light in a semiconductor sphere. Sov. Phys. Semicond. **16**, 772–775 (1982)
5. Ekimov, A.I., Onushchenko, A.A., Tsekhomskii, V.A.: Exciton absorption by copper(I) chloride crystals in a glassy matrix. Fizika i Khimiya Stekla **6**, 511–512 (1980)
6. Gates, B.D., Xu, Q., Stewart, M., Ryan, D., Willson, C.G., Whitesides, G.M.: New approaches to nanofabrication: Molding, printing, and other techniques. Chem. Rev. **105**, 1171–1196 (2005)
7. Gurav, A., Kodas, T., Pluym, T., Xiong, Y.: Aerosol processing of materials. Aerosol Sci. Technol. **19**, 411–452 (1993)
8. Henglein, A.: Photo-degradation and fluorescence of colloidal cadmium sulfide in aqueous solution. Ber. Bunsenges. Phys. Chem. **86**, 301–305 (1982)
9. Henglein, A.: Small particle research: Physicochemical properties of extremely small colloidal metal and semiconductor particles. Chem. Rev. **89**, 1861–1873 (1989)
10. Iijima, S.: Helical microtubules of graphitic carbon. Nature **354**, 56–58 (1991)

11. Kalyanasundaram, K., Borgarello, E., Duonghong, D., Grätzel, M.: Cleavage of water by visible-light irradiation of colloidal CdS solutions; inhibition of photocorrosion by RuO_2. Angew. Chem. Int. Ed. **20**, 987–988 (1981)
12. Kruis, F.E., Fissan, H., Peled, A.: Synthesis of nanoparticles in the gas phase for electronic, optical and magnetic applications: A review. J. Aerosol Sci. **29**, 511–535 (1998)
13. Lu, W., Lieber, C.M.: Nanoelectronics from the bottom up. Nature Mater. **6**, 841–850 (2007)
14. Matijevic, E.: Production of monodispersed colloidal particles. Annu. Rev. Mater. Sci. **15**, 483–516 (1985)
15. Matijevic, E.: Preparation and properties of uniform size colloids. Chem. Mater. **5**, 412–426 (1993)
16. Murray, C.B., Kagan, C.R., Bawendi, M.G.: Synthesis and characterization of monodisperse nanocrystals and close-packed nanocrystal assemblies. Annu. Rev. Mater. Sci. **30**, 545–610 (2000)
17. Murray, C.B., Norris, D.J., Bawendi, M.G.: Synthesis and characterization of nearly monodisperse CdE (E=S, Se, Te) semiconductor nanocrystallites. J. Am. Chem. Soc. **115**, 8706–8715 (1993)
18. Ozin, G.A.: Panoscopic materials: Synthesis over 'all' length scales. Chem. Commun. pp. 419–432 (2000)
19. Peng, X.G., Manna, L., Yang, W.D., Wickham, J., Scher, E., Kadavanich, A., Alivisatos, A.P.: Shape control of CdSe nanocrystals. Nature **404**, 59–61 (2000)
20. Rao, C.N.R.: Transition metal oxides. Annu. Rev. Phys. Chem. **40**, 291–326 (1989)
21. Rao, C.N.R., Raveau, B.: Transition metal oxides. VCH Publishers Inc., New York (1995)
22. Rogach, A.L., Talapin, D.V., Shevchenko, E.V., Kornowski, A., Haase, M., Weller, H.: Organization of matter on different size scales: Monodisperse nanocrystals and their superstructures. Adv. Funct. Mater. **12**, 653–664 (2002)
23. Rossetti, R., Ellison, J.L., Gibson, J.M., Brus, L.E.: Size effects in the excited electronic states of small colloidal CdS crystallites. J. Chem. Phys. **80**, 4464–4469 (1984)
24. Shevchenko, E.V., Talapin, D.V., Kotov, N.A., O'Brien, S., Murray, C.B.: Structural diversity in binary nanoparticle superlattices. Nature **439**, 55–59 (2006)
25. Strobel, R., Pratsinis, S.E.: Flame aerosol synthesis of smart nanostructured materials. J. Mater. Chem. **17**, 4743–4756 (2007)
26. Thompson, S.E., Parthasarathy, S.: Moore's law: The future of Si microelectronics. Materials Today **9**, 20–25 (2006)
27. Whitesides, G.M.: Nanoscience, nanotechnology, and chemistry. Small **1**, 172–179 (2005)
28. Whitesides, G.M., Grzybowski, B.: Self-assembly at all scales. Science **295**, 2418–2421 (2002)
29. Zhang, H., Edwards, E.W., Wang, D., Möhwald, H.: Directing the self-assembly of nanocrystals beyond colloidal crystallization. Phys. Chem. Chem. Phys. **8**, 3288–3299 (2006)

Chapter 2
Aqueous and Nonaqueous Sol-Gel Chemistry

2.1 Introduction

The most widely used synthetic technique for bulk metal oxides has been the ceramic method, which is based on the direct reaction of powder mixtures. These reactions are completely controlled by the diffusion of the atomic or ionic species through the reactants and products. To bring the reaction partners sufficiently close together and to provide high mobility, these solid state processes require high temperature and small particle sizes. Although the harsh reaction conditions only lead to thermodynamically stable phases, preventing the formation of metastable solids, these approaches gave access to a large number of new solid compounds, enabling the development of structure-properties relationships. However, in comparison to organic chemistry, where highly sophisticated synthetic pathways are employed to make and break chemical bonds in a controlled way, the ceramic method is a rather crude approach. It is therefore no surprise that for the size- and shape-controlled synthesis of nanoparticles especially liquid-phase routes represent the most promising alternatives. In contrast to solid-state processes, but analogous to organic chemistry, "chimie douce" approaches offer the possibility to control the reaction pathways on a molecular level during the transformation of the precursor species to the final product, enabling the synthesis of nanoparticles with well-defined and uniform crystal morphologies and with superior purity and homogeneity [12]. Among the various soft-chemistry routes, sol-gel procedures were particularly successful in the preparation of bulk metal oxides (e.g., ceramics, glasses, films and fibers) [33, 20], and therefore they have also been applied for nanoparticle synthesis. But in spite of great efforts, the number of oxidic nanoparticles obtained by sol-gel chemistry is still rather small compared to the variety of compounds obtained via powder routes. It turned out that in many cases a synthesis protocol developed for a bulk metal oxide could not directly be adapted to its corresponding counterpart on the nanoscale. The reasons for this observation are manifold. Aqueous sol-gel chemistry is quite complex, on the one hand due to the high reactivity of the metal oxide precursors towards water and the double role of water as ligand and solvent, and, on the other hand, due to the large number of reaction parameters that have to be strictly controlled (hydrolysis and condensation rate of the metal oxide precursors, pH, temperature, method of mixing, rate of oxidation, the nature and concentration of anions, ...) in order to provide good reproducibility of the synthesis protocol [33]. Another fundamental problem of aqueous sol-gel chemistry is that the as-synthesized precipitates are generally amorphous. The required post-synthetic annealing step to induce the crystallization process prevents any subtle control over crystal size and shape. For the preparation of bulk metal oxides these limitations play only a minor role, however, in the case of nanoparticle synthesis they constitute a major issue.

Nonaqueous (or non-hydrolytic) sol-gel processes in organic solvents, generally under exclusion of water, are able to overcome some of the major limitations of aqueous systems, and thus represent a powerful and versatile alternative [30, 40, 38, 37, 41]. The advantages are a direct consequence of the manifold role of the organic components in the reaction system (e.g., solvent, organic ligand of the precursor molecule, surfactants, or *in situ* formed organic condensation products). On the one hand they act as oxygen-supplier for the oxide formation and strongly determine the particle size and shape as well as the surface properties due to their coordination properties, and on the other hand the moderate reactivity of the oxygen carbon bond generally results in slower reaction rates. Altogether these parameters lead to the situation, where nonaqueous synthesis routes generally yield metal oxide nanoparticles with uniform, yet complex crystal morphologies, crystallite sizes in the range of just a few nanometers, and good dispersibility in organic solvents. Another important point is the fact that the chemistry of the oxygen-carbon bond is well-known from organic chemistry. This aspect is of utmost significance considering the fundamental role of organic reaction pathways in nonaqueous sol-gel chemistry approaches (cf. Chapter 5). Parallel to the formation of the inorganic nanoparticles, also the initial organic species undergo transformation reactions often following elementary mechanisms of organic chemistry [39]. Based on the identification and quantification of these organic by-products, it is possible to correlate the processes leading to these organic species by retro-synthetical analysis to the growth mechanisms of the oxide nanoparticles. The connection of organic chemistry with the synthesis of inorganic nanomaterials thus represents first, but fundamental steps towards a future rational synthesis design for inorganic nanoparticles.

We have to point out that many examples discussed in this book do not strictly follow the definition of sol-gel chemistry, i.e., the transformation of the molecular precursor does not exclusively proceed along the formation of a sol and then a gel. However, we denote any process as "sol-gel", as long as the transformation of the molecular precursor into the final oxidic compound involves chemical condensation reactions in liquid-phase under mild conditions. This definition allows us to draw comparisons between aqueous and nonaqueous sol-gel processes, highlighting the analogy between the mechanistic pathways found in both methods. We will also discuss some examples in this book that are not based on sol-gel chemistry, but on simple thermal decomposition of molecular precursors in liquid media. Furthermore, in the current literature the synthesis of metal oxides in organic solvents is typically denoted as “nonhydrolytic”. However, in some cases these procedures involve the use of hydrated metal oxide precursors and/or water is produced *in situ*, so that a hydrolytic reaction pathway cannot be excluded. Therefore, we prefer to name these routes as “nonaqueous sol-gel chemistry”.

2.2 Aqueous Sol-Gel Chemistry

The aqueous sol-gel process can shortly be defined as the conversion of a precursor solution into an inorganic solid via inorganic polymerization reactions induced by water. In general, the precursor or starting compound is either an inorganic (no carbon) metal salt (chloride, nitrate, sulfate, ...) or a metal organic compound such as an alkoxide. Metal alkoxides are the most widely used precursors, because they react readily with water and are known for many metals [47, 6]. Some alkoxides, which are widely used in industry, are commercially available at low cost (Si, Ti, Al, Zr), whereas other ones are hardly available, or only at very high costs (Mn, Fe, Co, Ni, Cu, Y, Nb, Ta, ...). In comparison to the preparation of silicates from silicon alkoxides, sol-gel processing of transition metal oxides has much less been studied, mainly due to the high reactivity of transition metal alkoxides.

In general, the sol-gel process consists of the following steps (Figure 2.1) [34]: i) Preparation of a homogeneous solution either by dissolution of metal organic precursors in an organic solvent that is miscible with water, or by dissolution of inorganic salts in water; ii) conversion of the homogeneous solution into a sol by treatment with a suitable reagent (generally water with or without any acid/base); iii) aging; iv) shaping; and v) thermal treatment/sintering. The first step in a sol-gel reaction is the formation of an inorganic polymer by hydrolysis and condensation reactions, i.e., the transformation of the molecular precursor into a highly crosslinked solid. Hydrolysis leads to a sol, a dispersion of colloidal particles in a liquid, and further condensation results in a gel, an interconnected, rigid and porous inorganic network enclosing a continuous liquid phase. This transformation is called the sol-gel transition. There are two possibilities to dry the gels. Upon removal of the pore liquid under hypercritical conditions, the network does not collapse and aerogels are produced. When the gel is dried under ambient conditions, shrinkage of the pores occurs, yielding a xerogel. One of the highly attractive features of the sol-gel process is the possibility to shape the material into any

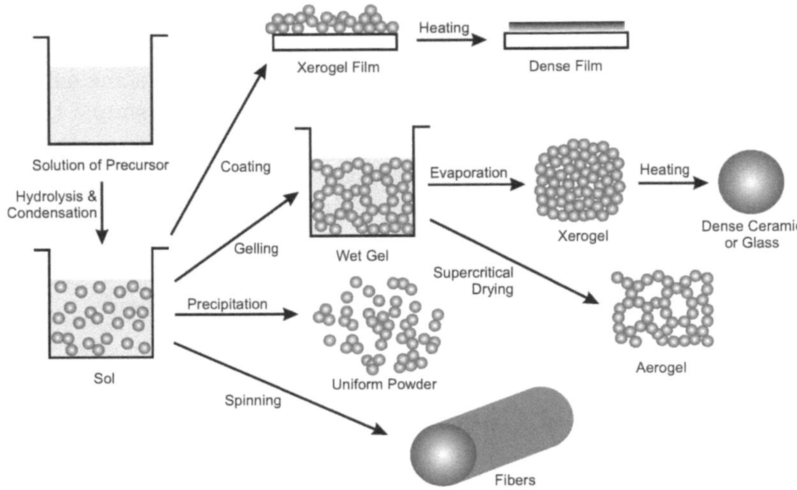

Fig. 2.1. Various steps in the sol-gel process to control the final morphology of the product

desired form such as monoliths, films, fibers, and monosized powders, and subsequently to convert it into a ceramic material by heat treatment.

As mentioned before, the sol-gel processes can be classified into two different routes depending on the nature of the precursors: a) the precursor is an aqueous solution of an inorganic salt or b) a metal organic compound [33]. The inorganic route involves the formation of condensed species from aqueous solutions of inorganic salts by adjusting the pH, by increasing the temperature or by changing the oxidation state. But this method has several disadvantages. The aqueous chemistry of transition metal ions can be rather complicated because of the formation of a large number of oligomeric species, depending on the oxidation state, the pH or the concentration. The role of the counter anions, which are able to coordinate the metal ion giving rise to a new molecular precursor with different chemical reactivity towards hydrolysis and condensation, is almost impossible to predict. These ions can influence the morphology, the structure and even the chemical composition of the resulting solid phase. Also the removal of these anions from the final metal oxide product is often a problem. Many of these issues can be avoided by using metal alkoxides as precursors. They are often soluble in organic solvents, providing high homogeneity, and they can easily be converted to the corresponding oxide.

The sol-gel conversion of metal alkoxides involves two main reaction types: hydrolysis and condensation (Scheme 2.1). During hydrolysis, the alkoxide groups (-OR) are replaced via the nucleophilic attack of the oxygen atom of a water molecule under release of alcohol and the formation of a metal hydroxide. Condensation reactions between two hydroxylated metal species leads to M-O-M bonds under release of water (oxolation), whereas the reaction between a hydroxide and an alkoxide leads to M-O-M bonds under release of an alcohol (alkoxolation).

$$\equiv M{-}OR + H_2O \longrightarrow \equiv M{-}OH + ROH \qquad (2.1)$$

$$\equiv M{-}OH + HO{-}M\equiv \longrightarrow \equiv M{-}O{-}M\equiv + H_2O \qquad (2.2)$$

$$\equiv M{-}OR + HO{-}M\equiv \longrightarrow \equiv M{-}O{-}M\equiv + ROH \qquad (2.3)$$

Scheme 2.1. Main reactions in the sol-gel process using metal alkoxides. Hydrolysis (Eq. 2.1) and condensation, involving oxolation (Eq. 2.2) and alkoxolation (Eq. 2.3)

Chemical aspects play an important role in studying and controlling the sol-gel process. The chemical reactivity of metal alkoxides towards hydrolysis and condensation depends mainly on the electronegativity of the metal atom, its ability to increase the coordination number, the steric hindrance of the alkoxy group, and on the molecular structure of the metal alkoxides (monomeric or oligomeric). The amount of added water in the hydrolysis step and how the water is added, determines, whether the alkoxides are completely hydrolyzed or not and which oligomeric intermediate species are formed. Additional parameters are the polarity, the dipole moment, and the acidity of the solvent.

The major problem of sol-gel methods based on the hydrolysis and condensation of molecular precursors is the control over the reaction rates. For most transition metal oxide precursors, these reactions are too fast, resulting in loss of morphological and also structural control over the final oxide material. Furthermore, the different reactivities of metal alkoxides make it difficult to control the composition and the homogeneity of complex multimetal oxides by the sol-gel process. One possibility to decrease and to adjust the reactivity of the precursors is the use of organic additives like carboxylic acids, β-diketones or functional alcohols, which act as chelating ligands and modify the reactivity of the precursors [33, 21]. An alternative strategy involves the slow release of water by chemical or physical processes, allowing control over the local water concentration and thus, over the hydrolysis of the metal oxide precursors [9]. In spite of all these efforts, the strong sensitivity of aqueous sol-gel processes towards any slight changes in the synthesis conditions and the simultaneous occurrence of hydrolysis and condensation reactions makes it still impossible to fully control the sol-gel processing of metal oxides in aqueous medium.

More information and mechanistic details about the aqueous sol-gel processing of metal oxides can be found in several excellent books and reviews [33, 20, 7, 29].

2.3 Nonaqueous Sol-Gel Chemistry

In nonaqueous sol-gel chemistry the transformation of the precursor takes place in an organic solvent under exclusion of water. In comparison to aqueous sol-gel chemistry, the list of potential precursors is longer and includes, in addition to inorganic metal salts and metal alkoxides, also metal acetates and metal acetylacetonates. Organometallic compounds are also frequently used, but then the process is rather based on thermal decomposition than sol-gel (remember: organometallic compounds are defined as having a direct metal-carbon bond, whereas metal organic compounds have a metal-hetero atom (usually oxygen) bond between the metal center and the organic ligand).

Nonaqueous sol-gel processes have a long history, although they never became as popular as their aqueous analogs. Early studies on nonaqueous sol-gel processes date back to the middle of the 19th century, when the reaction between various metal chlorides and alcohols were investigated. Ebelmen, for example, found that silicon tetrachloride forms silica gels in ethanol [14]. More than 80 years later, Dearing and Reid presented their work on "Alkyl Orthosilicates", proposing an aqueous and nonaqueous route to silica gels, assuming that they might show "different absorptive power" [13]. Many other groups continued to work on this topic [42, 17, 16], including peculiar but interesting examples like the conversion of chlorosilanes to oligosiloxanes and of silicon and titanium tetrachloride to the respective oxide by refluxing in dimethylformamide [43], the reaction of silicon tetrachloride with benzaldehyde [49] and organic ethers [45], or the synthesis of silica-sodalite from nonaqueous systems [4]. Nevertheless, these publications were more or less isolated efforts without receiving much attention. However, in the middle of the 1980s, and more intense around the beginning of the 1990s, research on nonhydrolytic preparation routes to metal oxides became popular to a larger scientific community [9, 48, 19]. Two main research directions could be distinguished: One focusing on the preparation of metal oxide gels, the other on metal oxide powders. Several groups worked on the preparation of powders using metal alkoxides and alcohols [15, 23, 25] or inert organic solvents [25, 26, 24]. Especially the glycothermal method involving the reaction of metal alkoxides or acetylacetonates with 1,4-butanediol is a versatile approach to various metal oxides [31] and has recently been reviewed [22]. On the other side, in 1992 Corriu et al. published their work on monolithic silica, alumina, and titania gels, as well as on gels containing two metals [8, 11, 10]. In the same year, the synthesis of zinc oxide gels was presented, using zinc alkoxides as precursors and acetone as condensation agent [18]. Upon aging, the gels formed a precipitate, which contained 4 nm zincite nanocrystals. An analogous approach was later on used for TiO_2, BaO, and $BaTiO_3$ [3]. Following these examples, a large variety of metal oxide gels were synthesized and reported, including silica, alumina, titania, mixed Al/Si and transition metal oxides from the corresponding metal chlorides and metal alkoxides or ethers as oxygen donors [1, 2], metal ferrites from the metal nitrates and ethylene glycol

[44], and zirconia and zircon from metal carboxylates and metal alkoxides [28]. Although some of these "earlier" examples also led to the formation of nanoscale materials, it was probably the work on the synthesis of titanium oxide nanoparticles, published in 1999 by three independent groups [46, 32, 27] that gave the starting signal for the intensified search of new synthesis routes to metal oxide nanoparticles under nonaqueous and/or nonhydrolytic reaction conditions. Nowadays, the family of metal oxide nanoparticles that are prepared by nonaqueous processes has grown immensely and ranges from simple binary metal oxides to more complex ternary, multi-metal and doped systems.

In aqueous systems, metal alkoxides are the most widely used precursors, and their chemical transformation into the oxidic network involves hydrolysis and condensation reactions. In aqueous sol-gel processes the oxygen for the formation of the oxidic compound is supplied by the water molecules. In nonaqueous systems, where intrinsically no water is present, the question arises, where the oxygen for the metal oxide comes from. Analogous to the nonhydrolytic preparation of bulk metal oxide gels [25], the oxygen for nanoparticle formation is provided by the solvent (ethers, alcohols, ketones or aldehydes) or by the organic constituent of the precursor (alkoxides or acetylacetonates). It is interesting to note that in spite of the large number of methodologies reported for the nonaqueous synthesis of metal oxide nanoparticles, and independent whether surfactants are used or not, most of the known condensation steps, i.e., the formation of the metal-oxygen-metal bond as basic structural unit, can be summarized in only five distinct pathways (Scheme 2.2) [37, 39]:

- Alkyl halide elimination (Eq. 2.4)
- Ether elimination (Eq. 2.5)
- Condensation of carboxylate groups (ester and amide eliminations) (Eq. 2.6)
- C-C coupling of benzylic alcohols and alkoxide molecules (Eq. 2.7)
- Aldol/ketimine condensation (Eq. 2.8).

Alkyl halide elimination, the condensation between metal halides and metal alkoxides (either initially added, or formed upon the reaction of metal halides with alcohols) under release of an alkyl halide, is shown in Eq. 2.4. Ether elimination (Eq. 2.5) is the result of the reaction between two metal alkoxides. The ester elimination process involves the reaction between metal carboxylates and metal alkoxides or between metal carboxylates and alcohols (Eq. 2.6). Analogous to ester eliminations are amide eliminations, e.g. reactions between metal carboxylates and amines. Alkyl halide, ether and ester eliminations are the most commonly reported routes. However, due to the excellent catalytic activity of the metal centres in the precursor species peculiar and more complex organic reactions such as C-C bond formation between alkoxy groups were also observed (Eq. 2.7). Whereas in some cases the presence of a basic species was a prerequisite for C-C bond formation, transition metals with high Lewis acidity, such as Nb, Y, and Ce, were able to directly catal-

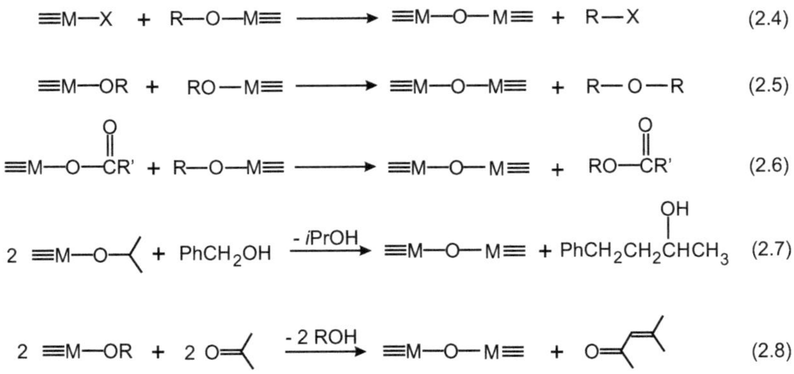

Scheme 2.2. Main condensation steps in nonaqueous sol-gel processes resulting in the formation of a metal-oxygen-metal bond. Alkyl halide elimination (Eq. 2.4), ether elimination (Eq. 2.5), ester elimination (Eq. 2.6), C-C bond formation between benzylic alcohols and alkoxides (Eq. 2.7), aldol condensation reactions (Eq. 2.8)

yse this Guerbet-like reaction. If ketones are used as solvents, the release of oxygen usually involves aldol condensation, where two carbonyl compounds react with each other under formal elimination of water (Eq. 2.8). The water molecules act as oxygen supplying agent for the metal oxide formation. However, all these condensation mechanisms are discussed in more details in Chapter 5.

2.4 Surfactant-Directed vs. Solvent-Controlled Nonaqueous Sol-Gel Approaches

Nonaqueous processes can roughly be divided into surfactant- and solvent-controlled approaches. Surfactant-controlled synthesis routes involve the transformation of the precursor species into the oxidic compound in the presence of stabilizing ligands in a typical temperature range of 250 to 350 °C. The most popular and successful approach within this strategy is the hot injection method, where the reagents are injected into a hot surfactant solution, which was particularly successfully applied in the synthesis of semiconductor nanocrystals [35].

An elegant alternative to surfactants is the use of common organic solvents, which act as reactant as well as control agent for particle growth, enabling the synthesis of high-purity nanomaterials in surfactant-free medium [41]. In comparison to the synthesis of metal oxides in the presence of surfactants the solvent-controlled approaches are simpler, because the initial reaction mixture just consists of two components, the metal oxide precursor(s) and a common organic solvent. The small number of reactants simplifies the characterization of the final reaction solution and, related to that, the elucidation of the chemical reaction mechanisms. The synthesis temperature is typically in the range of 50 to 200 °C, which is notably lower than in the hot-injection method.

A comparison of the literature on both approaches clearly shows that surfactant routes permit outstanding control over the growth of metal oxide nanoparticles, leading to almost perfectly monodisperse samples [40]. The ability of surfactants to cap the surface of the nanoparticles, sometimes combined with selectivity towards specific crystal faces, provides advantages such as shape control, low agglomeration tendency, good dispersibility in organic solvents, and the potential to tailor the surface properties. However, drawbacks resulting from surface-adsorbed surfactants are the unpredictable influence on the toxicity of the nanoparticles [36, 5], and the diminished accessibility of the particle surface, which is a serious issue regarding applications in gas sensing or catalysis. Although solvent-controlled approaches generally result in some agglomeration, the dispersibility properties of the nanoparticles can be improved by a post-synthetic functionalization step. In some cases a tiny amount of surfactants suffices to lead to completely transparent nanoparticle dispersions. However, surfactant-assisted and surfactant-free synthesis approaches both have advantages and limitations. Table 2.1 summarizes the pros and cons of the two strategies, clearly underlining that they are rather complementary [41]. Of course, one always has to be careful with such general trends, because selected reaction systems can behave in a completely opposite way.

Table 2.1. Pros and cons of surfactant-directed and solvent-controlled nonaqueous liquid-phase routes to metal oxide nanoparticles

	Surfactant-directed approaches	Solvent-controlled approaches
Pros		
	• Excellent control over crystal size • Narrow size distribution • Good control over crystal shape • Low agglomeration tendency • Good redispersibility	• Low amount of organic impurities • Non-toxic solvents • Simple, robust and widely applicable synthesis protocols • Good accessibility of the nanoparticle surface
Cons		
	• Large amount of organic impurities • Toxicity of surfactants • Restricted accessibility of the nanoparticle surface • Complex reaction mixtures	• Less control over crystallite size and shape • Broader size distributions • Formation of agglomerates • Restricted redispersibility

References

1. Acosta, S., Arnal, P., Corriu, R.J.P., Leclercq, D., Mutin, P.H., Vioux, A.: A general nonhydrolytic sol-gel route to oxides. Mater. Res. Soc. Symp. Proc. **346**, 43–64 (1994)
2. Arnal, P., Corriu, R.J.P., Leclercq, D., Mutin, P.H., Vioux, A.: Preparation of transition metal oxides by a nonhydrolytic sol-gel process. Mater. Res. Soc. Symp. Proc. **346**, 339–344 (1994)
3. Bates, S.E., Buhro, W.E.: Aldolate complexes as thermal precursors to metal-oxides - a new nonhydrolytic sol-gel strategy. Abstr. Pap. Am. Chem. Soc. **205**, 83INOR (1993)
4. Bibby, D.M., Dale, M.P.: Synthesis of silica-sodalite from non-aqueous systems. Nature **317**, 157–158 (1985)
5. Borm, P.J.A., Robbins, D., Haubold, S., Kuhlbusch, T., Fissan, H., Donaldson, K., Schins, R., Stone, V., Kreyling, W., Lademann, J., Krutmann, J., Warheit, D., Oberdorster, E.: The potential risk of nanomaterials: A review carried out for ECETOC. Part. Fibre Toxicol. **3**, 11–45 (2006)
6. Bradley, D.C., Mehrotra, R.C., Rothwell, I.P., Singh, A.: Alkoxo and aryloxo derivatives of metals. Academic Press: London (2001)
7. Brinker, C.J., Scherer, G.W.: Sol-gel science. Academic Press, San Diego (1990)
8. Corriu, R., Leclercq, D., Lefevre, P., Mutin, P.H., Vioux, A.: Preparation of monolithic binary oxide gels by a nonhydrolytic sol-gel process. Chem. Mater. **4**, 961–963 (1992)
9. Corriu, R.J.P., Leclercq, D.: Recent developments of molecular chemistry for sol-gel processes. Angew. Chem. Int. Ed. **35**, 1420–1436 (1996)
10. Corriu, R.J.P., Leclercq, D., Lefevre, P., Mutin, P.H., Vioux, A.: Preparation of monolithic gels from silicon halides by a non-hydrolytic sol-gel process. J. Non-Cryst. Solids **146**, 301–303 (1992)
11. Corriu, R.J.P., Leclercq, D., Lefevre, P., Mutin, P.H., Vioux, A.: Preparation of monolithic metal-oxide gels by a non-hydrolytic sol-gel process. J. Mater. Chem. **2**, 673–674 (1992)
12. Cushing, B.L., Kolesnichenko, V.L., O'Connor, C.J.: Recent advances in the liquid-phase syntheses of inorganic nanoparticles. Chem. Rev. **104**, 3893–3946 (2004)
13. Dearing, A.W., Reid, E.E.: Alkyl orthosilicates. J. Am. Chem. Soc. **50**, 3058–3062 (1928)
14. Ebelmen: Untersuchungen über die Verbindungen der Borsäure und Kieselsäure mit Aether. Annal. Chem. Pharm. **57**, 334 (1846)
15. Fanelli, A.J., Burlew, J.V.: Preparation of fine alumina powder in alcohol. J. Am. Ceram. Soc. **69**, C174–C175 (1986)
16. Gerrard, W., Kilburn, K.D.: Correlation between reactivity of the 1-carbon atom in alcohols, and certain properties of alkoxysilanes. J. Chem. Soc. pp. 1536–1539 (1956)
17. Gerrard, W., Woodhead, A.H.: Interaction of alcohols with silicon tetrachloride. J. Chem. Soc. pp. 519–522 (1951)
18. Goel, S.C., Chiang, M.Y., Gibbons, P.C., Buhro, W.E.: New chemistry for the sol-gel process: Acetone as a new condensation reagent. Mater. Res. Soc. Symp. Proc. **271**, 3–13 (1992)
19. Hay, J.N., Raval, H.M.: Synthesis of organic-inorganic hybrids via the non-hydrolytic sol-gel process. Chem. Mater. **13**, 3396–3403 (2001)
20. Hench, L.L., West, J.K.: The sol-gel process. Chem. Rev. **90**, 33–72 (1990)
21. Hubert-Pfalzgraf, L.G.: Some aspects of homo and heterometallic alkoxides based on functional alkohols. Coord. Chem. Rev. **178–180**, 967–997 (1998)

22. Inoue, M.: Glycothermal synthesis of metal oxides. J. Phys.: Condens. Matter **16**, S1291–S1303 (2004)
23. Inoue, M., Kitamura, K., Tanino, H., Nakayama, H., Inui, T.: Alcohothermal treatments of gibbsite: Mechanisms for the formation of boehmite. Clays Clay Miner. **37**, 71–80 (1989)
24. Inoue, M., Kominami, H., Inui, T.: Thermal transformation of χ-alumina formed by thermal decomposition of aluminum alkoxide in organic media. J. Am. Ceram. Soc. **75**, 2597–2598 (1992)
25. Inoue, M., Kominami, H., Inui, T.: Novel synthetic method for the catalytic use of thermally stable zirconia: Thermal decomposition of zirconium alkoxides in organic media. Appl. Catal., A **97**, L25–L30 (1993)
26. Inoue, M., Kominami, H., Otsu, H., Inui, T.: Synthesis of microcrystalline titania in organic media. Nippon Kagaku Kaishi pp. 1364–1366 (1991)
27. Ivanda, M., Music, S., Popovic, S., Gotic, M.: XRD, Raman and FT-IR spectroscopic observations of nanosized TiO_2 synthesized by the sol-gel method based on an esterification reaction. J. Mol. Struct. **481**, 645–649 (1999)
28. Jansen, M., Guenther, E.: Oxide gels and ceramics prepared by a nonhydrolytic sol-gel process. Chem. Mater. **7**, 2110–2114 (1995)
29. Jolivet, J.P.: Metal oxide chemistry and synthesis. John Wiley & Sons Ltd.: Chichester, England (2000)
30. Jun, Y.W., Choi, J.S., Cheon, J.: Shape control of semiconductor and metal oxide nanocrystals through nonhydrolytic colloidal routes. Angew. Chem. Int. Ed. **45**, 3414–3439 (2006)
31. Kominami, H., Inoue, M., Inui, T.: Formation of niobium double oxides by the glycothermal method. Catal. Today **16**, 309–317 (1993)
32. Kominami, H., Kato, J., Murakami, S., Kera, Y., Inoue, M., Inui, T., Ohtani, B.: Synthesis of titanium IV oxide of ultra-high photocatalytic activity: High-temperature hydrolysis of titanium alkoxides with water liberated homogeneously from solvent alcohols. J. Mol. Catal. A: Chem. **144**, 165–171 (1999)
33. Livage, J., Henry, M., Sanchez, C.: Sol-gel chemistry of transition metal oxides. Prog. Solid State Chem. **18**, 259–341 (1988)
34. Mehrotra, R.C., Singh, A.: Recent trends in metal alkoxide chemistry. Prog. Inorg. Chem. **46**, 239–454 (1997)
35. de Mello Donega, C., Liljeroth, P., Vanmaekelbergh, D.: Physicochemical evaluation of the hot-injection method, a synthesis route for monodisperse nanocrystals. Small **1**, 1152–1162 (2005)
36. Nel, A., Xia, T., Mädler, L., Li, N.: Toxic potential of materials at the nanolevel. Science **311**, 622–627 (2006)
37. Niederberger, M.: Nonaqueous sol-gel routes to metal oxide nanoparticles. Acc. Chem. Res. **40**, 793–800 (2007)
38. Niederberger, M., Antonietti, M.: Nanomaterials chemistry: Recent developments and new directions, chap. Nonaqueous sol-gel routes to nanocrystalline metal oxides, pp. 119–138. Wiley-VCH (2007)
39. Niederberger, M., Garnweitner, G.: Organic reaction pathways in the nonaqueous synthesis of metal oxide nanoparticles. Chem. Eur. J. **12**, 7282–7302 (2006)
40. Park, J., Joo, J., Kwon, S.G., Jang, Y., Hyeon, T.: Synthesis of monodisperse spherical nanocrystals. Angew. Chem. Int. Ed. **46**, 4630–4660 (2007)
41. Pinna, N., Niederberger, M.: Surfactant-free nonaqueous synthesis of metal oxide nanostructures. Angew. Chem. Int. Ed. **47**, 5292–5304 (2008)
42. Ridge, D., Todd, M.: Studies in the formation mechanisms of alkyl orthosilicates. J. Chem. Soc. pp. 2637–2640 (1949)
43. Rochow, E.G., Gingold, K.: The conversion of chlorosilanes to siloxanes by dimethylformamide. J. Am. Chem. Soc. **76**, 4852–4855 (1954)
44. Schleich, D.M., Zhang, Y.: Preparation of some metal ferrite MFe_2O_4 thin films through a nonaqueous sol method. Mater. Res. Bull. **30**, 447–452 (1995)

45. Schwarz, R., Kuchen, W.: Über die Ätherspaltung mit Siliziumtetrachlorid. Chem. Ber. **89**, 169–178 (1956)
46. Trentler, T.J., Denler, T.E., Bertone, J.F., Agrawal, A., Colvin, V.L.: Synthesis of TiO_2 nanocrystals by nonhydrolytic solution-based reactions. J. Am. Chem. Soc. **121**, 1613–1614 (1999)
47. Turova, N.Y., Turevskaya, E.P.: The chemistry of metal alkoxides. Kluwer Academic Publishers, Boston (2002)
48. Vioux, A.: Nonhydrolytic sol-gel routes to oxides. Chem. Mater. **9**, 2292–2299 (1997)
49. Zappel, A.: The reaction of chlorosilanes with benzaldehyde. J. Am. Chem. Soc. **77**, 4228 (1955)

Chapter 3
Surfactant-Assisted Synthesis

3.1 Hot-Injection Method

In 1993, Murray et al. published the synthesis of monodisperse CdX (X=S, Se, Te) nanocrystallites in molten trioctylphosphine oxide (TOPO) [90]. This work provided the basis for the so-called hot-injection method, which involves the injection of a room-temperature ("cold") solution of precursor molecules into a hot solvent in the presence of surfactants [87] (Figure 3.1).

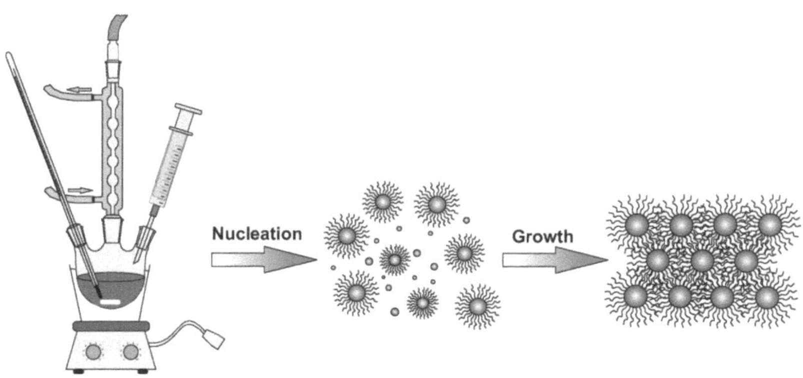

Fig. 3.1. Experimental set-up and reaction scheme for the hot-injection method

The fast injection of the precursor induces a high degree of supersaturation, resulting in a short burst of nucleation. During the nucleation process the precursor concentration in the solution decreases abruptly. The drop in temperature, due to the injection of the "cold" reactants, and the low concentration of unreacted remaining precursor prevent any further nucleation events. In a next step, the temperature is carefully increased to a value which allows the slow growth of the nuclei to larger nanoparticles, however still suppresses further nucleation. The separation of nucleation and growth is a prerequisite to synthesizing monodispersed nanoparticles [99], and therefore the hot-injection method generally leads to nanoparticles with a narrow size distribution of $\sigma \sim 10\%$. If size distributions narrower than 5% are required, then a size-selection process has to be applied.

The surfactants, typically consisting of a coordinating head group and a long alkyl chain, adsorb reversibly to the surfaces of the growing nanoparticles, and thus provide a dynamic organic capping layer that stabilizes the nanoparticles in solution and also mediates their growth [10]. The use of surfactants with selectivity towards specific crystal faces, or mixtures of surfactants with different binding affinities to the nanocrystal surface allow excellent control over crystal size, size distribution and morphology [71].

Finally, the nanocrystals can be precipitated by adding a nonsolvent. After their separation from the reaction liquid, the nanoparticles can be redispersed in suitable, usually apolar, organic solvents, forming stable colloidal suspensions. The surfactants used during the synthesis and now attached to the surface of the nanocrystals can be exchanged against other ones in a post-synthetic step, allowing not only the chemical modification of the surface properties of the nanoparticles in general, but also the tailoring of the dispersibility behavior in different solvents in particular [75].

The hot-injection method proved to be particularly versatile for preparing II-VI, IV-VI and III-V semiconductor nanocrystals [87, 99], however, it has also been extended to the synthesis of other materials such as metals and metal oxides. In 1999, Trentler et al. adapted the hot-injection method for the preparation of titania nanocrystals [140]. The procedure involved the injection of a titanium alkoxide into a hot solution of titanium tetrachloride, trioctylphosphine oxide and heptadecane. The off-white precipitate consisted of nearly spherical anatase nanoparticles with crystal sizes of less than 10 nm (Figure 3.2) and with excellent crystallinity (Figure 3.2, inset). Control over the crystal shape can be achieved upon addition of the surface-selective surfactant lauric acid (LA, $CH_3(CH_2)_{10}COOH$) in addition to TOPO as non-selective surfactant [60]. Depending on the LA concentration, the shape of the anatase nanocrystals varies from bullet- and diamond-like to rods and branched rods (Figure 3.3). The authors proposed that the growth of the nanocrystals is determined by the minimization of the surface area associated with the high energy facet [60].

In addition to titania, the hot-injection method was reported for the preparation of other metal oxides, e.g., MnO (injection of a manganese Cupferron complex dissolved in octylamine into trioctylamine at 300°C) [108], ZnO (injection of diethylzinc in decane into TOPO at 200°C) [121], ZrO_2 (injection of zirconium isopropoxide into a solution of $ZrCl_4$ and TOPO in heptadecane) [39], CeO_2 (injection of cerium benzoylacetonate dissolved in oleylamine into oleylamine at 250°C) [122], or γ-Fe_2O_3 (co-injection of iron pentacarbonyl in octyl ether and meta-chloro peroxybenzoic acid in octyl ether into a solution of tridecanoic acid in octyl ether at 293°C [14], or injection of iron pentacarbonyl into dodecylamine at 180°C [18].

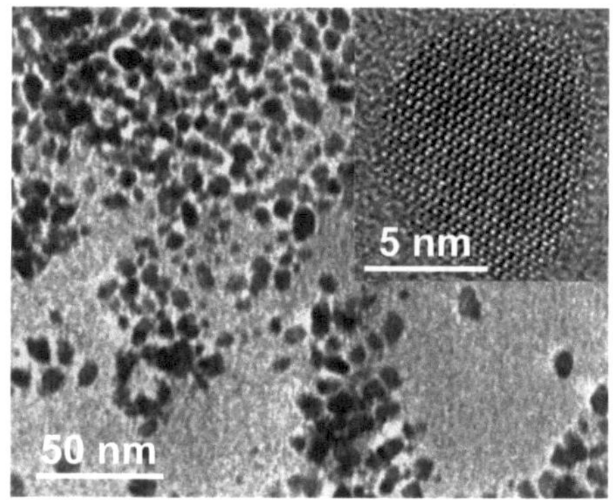

Fig. 3.2 TEM and HRTEM images of anatase nanoparticles obtained by the hot-injection method in TOPO. Image reproduced in part from Ref. [140] with permission of the American Chemical Society

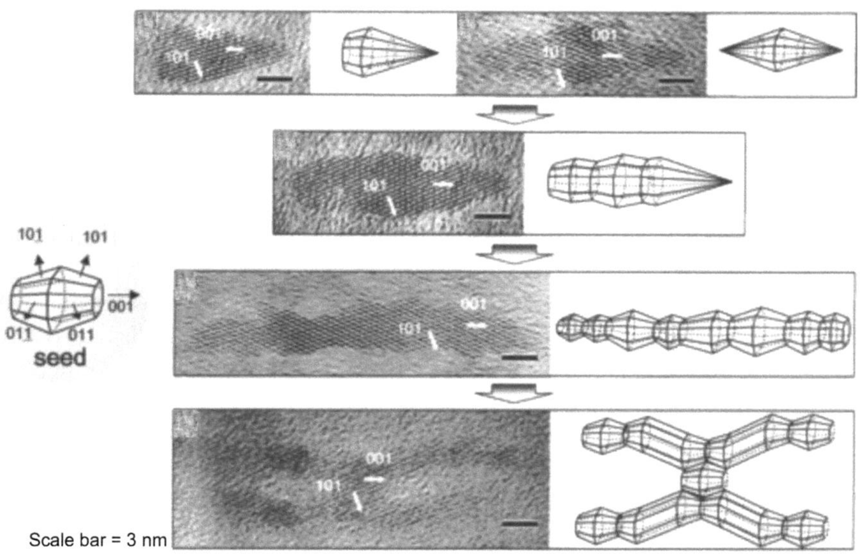

Fig. 3.3. HRTEM images and simulated three-dimensional shapes (bullet, diamond, short rod, long rod, and branched rod) that can be found for anatase nanocrystals synthesized by the hot-injection method using LA and TOPO as surfactants. Image reproduced in part from Ref. [60] with permission of the American Chemical Society

In addition to the injection of molecular precursors, also the rapid addition of metal oxide sols, amorphous nanoparticulate precursors, or reactive solvents to the hot solvent was reported as interesting modifications of the "standard" hot-injection method. In the first case, metal oxide sols were prepared by aqueous sol-gel techniques and subsequently injected into a solution of dodecylamine in tetradecene at 160°C [32]. The inorganic polymerization during the preparation of the sols was controlled through the addition of acetylacetone. The choice of the amine as surfactant was based on its ability to exchange the acetylacetonate ligands as well as to promote the condensation of the M-OH bonds on the particle surfaces. On the other hand, the amine acts as steric stabilizing agent, and consequently it is the competition between the different roles of the amine that determines the final size and structure of the products. The as-synthesized nanoparticles were either nanocrystalline (SnO_2, ZnO) or nearly amorphous (TiO_2, In_2O_3).

Another adaptation of the hot-injection method was reported for $BaTiO_3$ nanocrystals, involving first the synthesis of "pseudo" bimetallic precursors and their subsequent combination with a mixture of hot oleylamine and different surface coordinating ligands for crystallization and crystal growth [17]. The preparation of the bimetallic precursor was based on the dissolution of barium metal in benzyl alcohol, followed by the addition of titanium isopropoxide, similar to the procedure reported by Niederberger et al. [93, 94]. The formed white and amorphous precipitate, still dispersed in benzyl alcohol, was injected into a preheated (320°C) solvent/ligand mixture consisting of oleylamine/decanoic acid or oleylamine/oleic acid. Depending on the composition of the solvent/ligand mixture $BaTiO_3$ nanocrystals with diameters of 2–3, 3–5, or 6–10 nm were obtained.

The injection of a reactive solvent constitutes another strategy in the hot-injection method. Tang et al. applied this approach for the preparation of titania nanoparticles [136]. Bis(cyclooctatetraene)titanium $Ti(COT)_2$ was dissolved in anhydrous o-dichlorobenzene in the presence of coordinating ligands such as tributylphosphine, tributylphosphine oxide or TOPO. The injection of dimethyl sulfoxide (DMSO) to this mixture at 120°C instantaneously leads to a reaction of $Ti(COT)_2$ with DMSO, yielding monodispersed titania nanocrystals of about 15 nm and $S(CH_3)_2$ as organic side product. An analogous strategy was pursued by Zhang et al. for the preparation of titania nanorods with tunable aspect ratio [175]. In this case, titanium isopropoxide and oleic acid were heated in 1-octadecene, resulting in the formation of titanium carboxylate complexes. After heating the mixture to 260°C, oleylamine was injected, initiating the aminolysis reaction (cf. Chapter 5.4) that finally produced the titania nanorods. The length of the titania nanorods was easily tunable from 12 to 30 nm (Figure 3.4a,b), all with a diameter of about 2 nm, through the amount of oleylamine added. Ceria nanocrystals were obtained from cerium nitrate hexahydrate after the injection of diphenyl ether in the presence of various surfactants [161]. Depending on the surfactants and their relative ratio (oleylamine, oleylamine/tri-n-octylamine, or oleylamine/oleic acid) the shape of the ceria nanocrystals could be varied from spheres to nanowires and tadpole-shaped nanowires (Figure 3.4c). Other examples making use of the same strategy are the preparation of In_2O_3, ZnO, CoO, and MnO nanoparticles or nanoflowers from the corresponding metal acetates in the presence of myristic acid or stearic acid [92]. After dissolving the precursors and acids in 1-octadecene, the reaction mixture was heated to 280°C. Injection of alcohols such as 1-octadecyl alcohol or decyl alcohol as activation agents initiated the formation of the corresponding metal oxide nanomaterials. By reducing the degree of ligand protection it was possible to change the morphological characteristics of the products from stabilized primary nanocrystals to 3D nanostructures grown through oriented attachment (cf. also Chapter 6.2) [91].

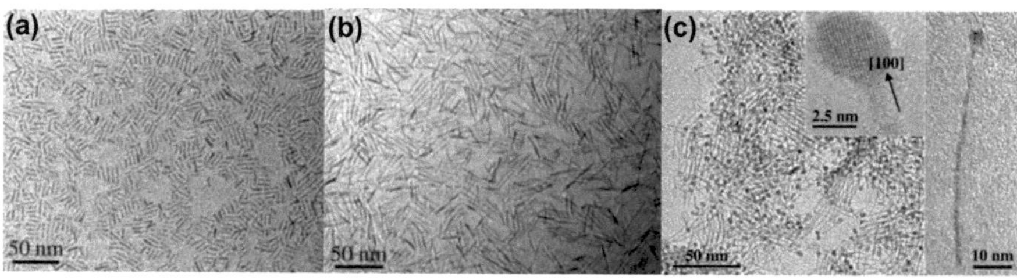

Fig. 3.4. a–c TEM overview images of titania nanorods with lengths of **a** 12 and **b** 30 nm. Reproduced partly from Ref. [175] with permission of Wiley-VCH. **c** TEM images of tadpole-shaped ceria nanocrystals (left: overview, right: one individual nanocrystal; inset: HRTEM image). Reproduced partly from Ref. [161] with permission of Wiley-VCH

3.2 Heating-Up Method

In the so-called "heating-up" method the reaction solution is prepared at low temperature. Subsequent heating initiates the crystallization process, which finally leads to the formation of nanocrystals. The heating-up method is particularly advantageous for large-scale production, because of its simplicity [99]. Nevertheless, the size uniformity that can be achieved by this approach is comparable to the best results obtained from the hot-injection method.

The heating-up method for the synthesis of magnetic iron oxide nanoparticles was pioneered by Hyeon et al. [50]. Adapting a synthesis protocol previously published by Bentzon et al. for the preparation of iron oxide superlattices [6], the procedure involved the injection of iron pentacarbonyl $Fe(CO)_5$ into a mixture of octyl ether and oleic acid at the moderate temperature of 100°C, producing iron oleate complexes as precursor species, whose exact structure is however unknown. During the subsequent heating to reflux (around 300°C) the iron oleate complexes decompose and metallic iron nanoparticles are generated that can easily be oxidized to γ-Fe_2O_3 by the mild oxidant trimethylamine oxide. The nanocrystals are highly monodisperse with diameters of 11 nm. Interestingly, it was not possible to increase the crystal diameter just by varying the $Fe(CO)_5$-to-oleic acid ratio. Instead, larger crystals were obtained by adding more iron oleate complex to the previously prepared 11 nm sized iron nanocrystallites. This seed-mediated growth is the most apparent strategy for the separation of nucleation and growth and has successfully been applied for the size-controlled synthesis of many different metal oxide nanoparticles (cf. also Section 3.6).

The heating-up method became the method of choice for magnetic nanoparticles such as spinel ferrites [49]. Figure 3.5 gives an overview of selected magnetic nanoparticles synthesized by this technique. Figure 3.5a and b display $MnFe_2O_4$ nanoparticles either with spherical or cubelike morphology. The synthesis involved the reaction of iron acetylacetonate and manganese acetylacetonate with 1,2-hexadecanediol, oleic acid, and oleylamine in benzyl ether [166]. Whereas the size was tuned by varying the concentration of the precursors, the shape was controlled by the amount of stabilizers added to the reaction mixture. At lower surfactant concentrations, the nanocrystals were spherical, with diameters of 12 nm (Figure 3.5a). Increasing the amount of surfactants yielded cubelike particles with basically the same size (Figure 3.5b). Manganese ferrite nanoparticles were also prepared from $Mn_2(CO)_{10}$ and $Fe(CO)_5$ in octyl ether and oleic acid [63]. The crystal size was adjusted by changing the molar ratio of precursors to oleic acid. A 1:1 ratio led to 5 nm-sized crystals (Figure 3.5c), whereas a ratio of 1:3 resulted in larger particles of 10 nm (Figure 3.5d), both with a spherical morphology. Uniform Mn_3O_4 nanoparticles were obtained upon the reaction of manganese acetylacetonate with oleylamine [115]. Figure 3.5e shows a TEM overview image of nearly monodisperse and spherical nanoparticles of 10 nm in diameter. The clearly visible lattice fringes in the HRTEM image prove the highly crystalline nature of these nanoparticles (Figure 3.5f). In contrast to many other colloidal routes the heating-up method has proven to be scalable, i.e., it enables the synthesis of several tens of grams of monodisperse nanoparticles in one batch without a size-sorting process [98]. Figure 3.6 shows an overall scheme leading to the formation of magnetite Fe_3O_4 nanocrystals, up to 40 g and larger than 95% in yields. The heating-up method constitutes nowadays one of the most popular synthesis routes and consequently, a large number of metal oxide nanoparticles have been prepared and reported based on this approach [99].

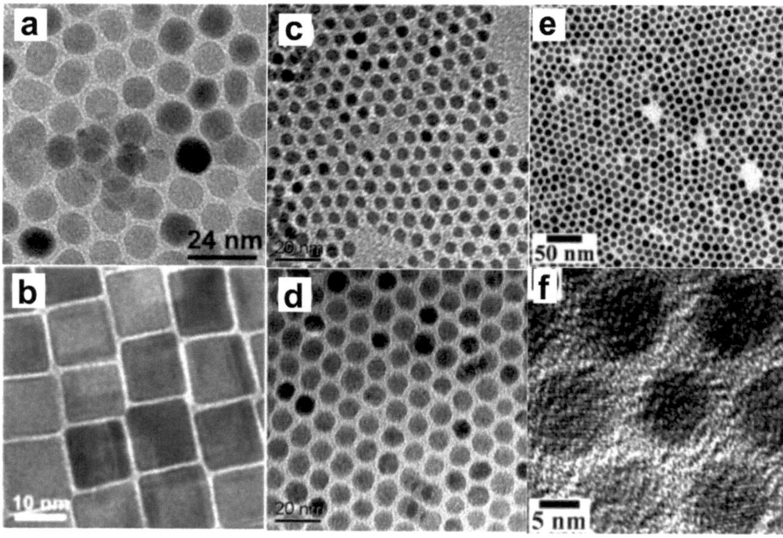

Fig. 3.5. **a–f** TEM images of magnetic nanoparticles: **a** spherical 12 nm sized $MnFe_2O_4$, **b** cubelike $MnFe_2O_4$, **c** spherical 5 nm-sized $MnFe_2O_4$, **d** spherical 10 nm-sized $MnFe_2O_4$, **e** spherical 10 nm-sized Mn_3O_4, and **f** HRTEM image of the nanoparticles displayed in **e**. Images **a** and **b** reproduced from Ref. [166], **c** and **d** from Ref. [63] with permission of the American Chemical Society. Images **e** and **f** reproduced from Ref. [115] with permission of Wiley-VCH

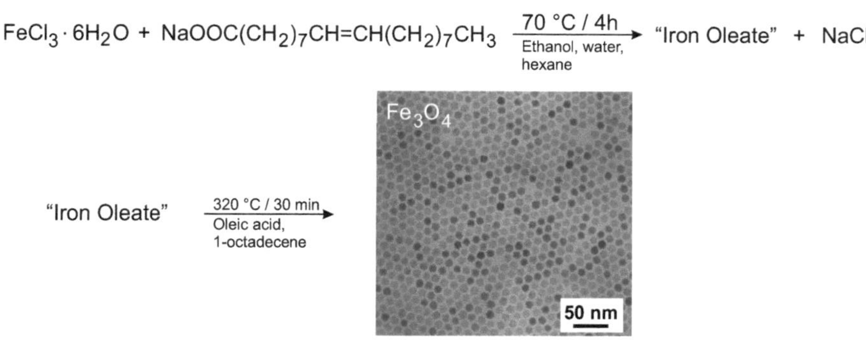

Fig. 3.6. Schematic overview of the synthesis pathway to monodisperse Fe_3O_4 nanocrystals involving the formation of iron oleate and its subsequent thermal decomposition

3.3 Comparison of the Heating-Up and Hot-Injection Method

Regarding the nucleation process, the main difference between the heating-up and the hot-injection method is the instantaneous supersaturation that is induced in the hot-injection method. Although the crystallization mechanisms underlying the control of the size distribution in the heating-up method are much less understood than in the hot-injection method, one would expect different growth kinetics for both methods. However, investigations in this direction did not yet show any significant differences, at least in comparison with other, obviously more important reaction parameters. In the case of ZnO nanoparticles for example, prepared by the thermal decomposition of zinc acetate in alkylamines in the presence of tert-butylphosphonic acid (TBPA), the UV-vis absorption spectra showed similar optical features for the final products prepared by the hot-injection and heating-up method, respectively [24]. On the other hand, it was found that TBPA was essential to keep the crystal size in the nanoscopic regime, and that the TBPA-to-zinc acetate ratio strongly influenced both the growth kinetics and the final crystal size. Comparison of the two methods was also drawn in the case of more complex morphologies like tetrapod-shaped maghemite nanocrystals [28]. The heating-up method involved the addition of iron pentacarbonyl, dissolved in 1-octadecene, to the three surfactants oleic acid, oleylamine and hexadecane-1,2-diol in 1-octadecene at 100°C, followed by heating to 240°C for 1h. The hot-injection procedure was different in that the iron pentacarbonyl was rapidly injected into the hot surfactant mixture at 240°C. However, both methods resulted in the formation of rather similar γ-Fe_2O_3 tetrapods, with the only difference that the arms of the tetrapods obtained by the hot-injection are slightly thicker. In both methods the main parameter determining the length of the arms of the tetrapods is the concentration of the precursor.

3.4 Solvothermal Synthesis

Solvothermal processes refer to chemical reactions that are performed in a closed reaction vessel at temperatures higher than the boiling point of the solvent employed. Solvothermal processes are mainly defined by chemical parameters such as the nature of the reagents and of the solvents, and by thermodynamical factors, in particular temperature and pressure [29]. Although solvothermal techniques have a long history, for a long time their whole focus was on the processing of bulk crystals and bulk materials, and fine grained powders, often obtained by these approaches, were discarded due to the lack of suitable characterization tools [11]. With the growing interest in sub-micron and nanosized powders with well-defined sizes and morphologies, solvothermal syntheses became a fundamental branch of nanoparticle preparation and processing. The trend towards greener technologies further strengthens these approaches, because on the one hand they consume less energy, due to the moderate reaction temperatures, and on the other hand they allow the use of common, inexpensive laboratory solvents, whose low boiling points constituted the main limitation for nanoparticle synthesis [106]. More information about hydro- and solvothermal processes including equipment and technical details can be found somewhere else [11]. Examples of metal oxide nanoparticles prepared under solvothermal conditions and in the presence of surfactants are magnetite [45, 46], ceria [130], hematite [82], tin oxide [149], cobalt oxides [155, 152, 83], or ZnO, N-doped ZnO and CdO [141]. An interesting example is the synthesis of anatase nanoparticles in toluene in the presence of oleic acid as surfactant [65]. In this case, the particle shape can be varied from nearly spherical (Figure 3.7a) to elongated, dumbbell-like nanorods (Figure 3.7b), either by increasing the concentration of titanium isopropoxide in toluene, or by increasing the concentration of oleic acid with respect to the precursor.

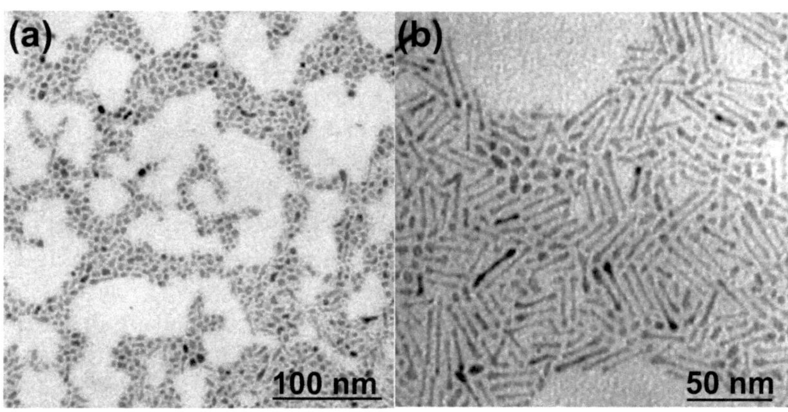

Fig. 3.7. **a**–**b** TEM overview images of titania nanoparticles prepared by varying the titanium isopropoxide-to-toluene weight ratios in the presence of a constant amount of oleic acid. **a** 5:100 and **b** 20:100. Images reproduced from Ref. [65] with permission of Elsevier Inc

3.5 Microwave Technique

The use of microwave irradiation in the synthesis of inorganic nanomaterials represents a rather new development, although this technique is well known in organic synthetic chemistry for more than two decades [40, 64, 129]. Microwave reactors operate at a frequency of 2.45 GHz, corresponding to a photon energy of 0.0016 eV, which is too low to break chemical bonds and which is also lower than the energy of Brownian motion. Consequently, microwaves are not able to induce chemical reactions, but they offer a very efficient heating tool. Irradiation of a reaction mixture at microwave frequencies results in alignment of the dipoles or ions in the electric field. Due to the fact that the applied field oscillates, the dipole or ion field has to realign itself continuously, which produces energy in the form of heat through molecular friction and dielectric loss [64]. The efficiency with which electromagnetic radiation is converted into heat is dependent on the dielectric properties of a solvent, i.e., solvents with strong ability to be polarized by the electric field are particularly advantageous for rapid heating. Organic solvents with high microwave absorbing properties are ethylene glycol, ethanol, dimetylsulfoxide (DMSO), isopropanol or formic acid [40]. Solvents without a permanent dipole moment like carbon tetrachloride, benzene or dioxane are nearly microwave transparent. Nevertheless, also these solvents can be used for microwave chemistry, as long as other reagents in the reaction mixture are polar. Furthermore, polar additives like ionic liquids can be added to otherwise low-absorbing media to increase the absorbance.

Fast and efficient heating as provided by microwave irradiation bears great potential for large-scale synthesis without suffering thermal gradient effects, and therefore this method became also interesting for the preparation of inorganic nanomaterials [33, 79]. The fast energy transfer directly to the reactants allows an instantaneous decomposition of nanocrystal precursors, creating highly supersaturated solutions. By varying the microwave irradiation time and the concentration of the surfactants it was possible to obtain rare earth oxide nanostructures ranging from small spherical nuclei to short rods and extended assemblies of nanowires [96]. In a typical reaction, metal acetates or metal acetylacetonates were dissolved in a mixture of oleic acid and oleylamine. After irradiating in a conventional domestic microwave oven for 5 to 15 minutes, uniform nanorods, nanowires and nanoplates of M_2O_3 (M = Pr, Nd, Sm, Eu, Gd, Tb, Dy) were obtained [96]. Figure 3.8 gives an overview of the as-synthesized Sm_2O_3 (Figure 3.8a) Nd_2O_3 (Figure 3.8b) and Gd_2O_3 (Figure 3.8c) nanorods and nanowires. The nanorods with an average diameter of 1.2 nm and an average length of 4–5 nm are self-assembled into 2D superstructures. The HRTEM image in Figure 3.8d clearly shows the crystalline nature of the Gd_2O_3 nanowires grown along the [100] direction. The authors also performed reference experiments, however using conventional heating and they found that the resulting nanorods were significantly less uniform than those obtained under microwave irradiation. From these findings they concluded that microwave chemistry offered better control of the morphology.

Another example for microwave-assisted processes is the synthesis of anatase nanocrystals in 1-butyl-3-methylimidazolium tetrafluoroborate as

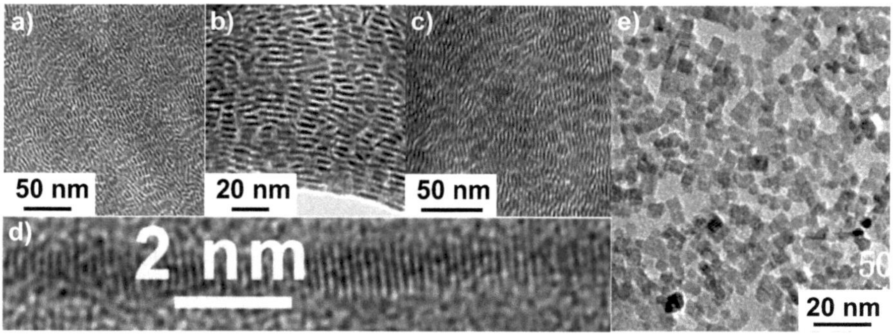

Fig. 3.8. **a–e** TEM images of **a** Sm_2O_3, **b** Nd_2O_3, **c** Gd_2O_3 nanorod assemblies, **d** HRTEM image of one Gd_2O_3 nanowire. **e** TEM overview images of anatase nanocrystals. Images **a–d** reproduced from Ref. [96], **e** from Ref. [30] with permission of the American Chemical Society

ionic liquid solvent and titanium isopropoxide as precursor [30]. After an irradiation time of 3–40 minutes, a milky suspension was obtained. TEM studies showed that the nanoparticles exhibited a uniform and cube-like shape with edge length of about 10 nm (Figure 3.8e). However, we have to point out that in this example titanium isopropoxide was hydrolyzed by traces of water present in the ionic liquid.

3.6 Seed-Mediated Growth

The seed-mediated growth is the most apparent approach for the separation of nucleation and growth. Nucleation is physically separated from growth by using preformed nanocrystals as seed nuclei, i.e., heterogeneous nucleation prevents the formation of additional nuclei by homogeneous nucleation. Preformed nuclei in the reaction solution are supplied with monomers that react on the surface of the nuclei. It is important that the monomer concentration is low enough to suppress homogeneous nucleation. Seed-mediated growth can be divided into two categories related to homogeneous and heterogeneous particles. Homogeneous particles consist of just one material, whereas heterogeneous particles have compositionally different sections, as for example found in core-shell structures. The latter are discussed in Chapter 3.8.

Seed-mediated growth has mainly been developed for the synthesis of metallic nanoparticles [53, 160], however, can also be used for metal oxides, in particular for magnetic nanoparticles. It is possible to achieve exceptionally narrow size distributions (standard deviation $\sigma < 5\%$), if the seed nuclei are also very uniform. An impressive example along these lines is the preparation of iron oxide nanocrystals, whose diameter can be adjusted in one nanometer increments [102]. The synthesis involves a two step process. Firstly, monodisperse iron nanoparticles were prepared from iron pentacarbonyl $Fe(CO)_5$ and oleic acid in dioctyl ether [49]. Depending on the molar $Fe(CO)_5$-to-oleic acid ratio the size of the iron nanoparticles can be adjusted from 4 to 8 and 11 nm. A very subtle size-tuning in one nanometer increments is then possible by reacting these seed nanoparticles with iron oleate solutions of defined concentrations, yielding monodisperse iron nanoparticles that transform into iron oxide nanoparticles of 6, 7, 8, 9 (Figure 3.9a–d), 10, 11, 12, and 13 nm on exposure to air [102]. The small size distribution leads to a hexagonally ordered arrangement of the nanoparticles on the TEM grid. Also spinel cobalt ferrite nanocrystals were obtained by a seed-mediated growth process [128]. In a first step, spherical $CoFe_2O_4$ nanocrystals with a diameter of 5 nm were prepared starting from cobalt and iron acetylacetonate in phenyl ether, 1,2-hexadecanediol, oleic acid and oleylamine (Figure 3.9e). These primary nanocrystals were used as seeds in additional precursor solutions to grow larger particles, either spherical or cube-shaped ones of 11 nm (Figure 3.9f). Similarly, $CoFe_2O_4$ (Figure 3.9g) and $MnFe_2O_4$ (Figure 3.9h) with crystal sizes of 14 nm were prepared [132].

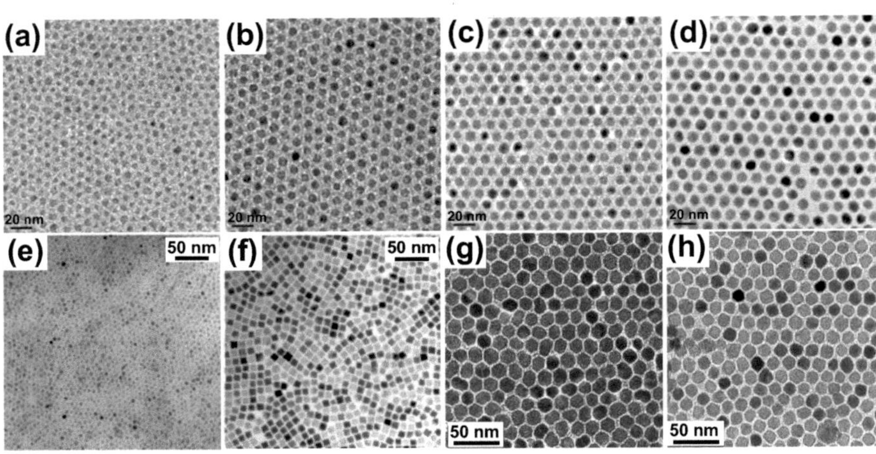

Fig. 3.9. **a–h** TEM overview images of iron oxide nanoparticles with diameters of **a** 6, **b** 7, **c** 8, **d** 9 nm proving the one nanometer size increments. Images reproduced from Ref. [102] with permission of Wiley-VCH. TEM overview images of **e** 5 nm-sized spherical $CoFe_2O_4$ nanocrystal seeds and **f** 11 nm-sized cube-shaped $CoFe_2O_4$ nanocrystals after seed-mediated growth. TEM overview images of 14 nm-sized **g** $CoFe_2O_4$ and h) $MnFe_2O_4$ nanoparticles made by seed-mediated growth. Images **e** and **f** reproduced from Ref. [128], images **g** and **h** from Ref. [132] with permission of the American Chemical Society

3.7 Self-Assembled Nanoparticles

Surfactants as amphiphilic molecules are characterized by a rich self-assembly chemistry and therefore they offer the possibility to grow and assemble the nanoparticles in one step. In these cases, the surfactants not only influence the size and shape of the nanoparticles, but also arrange them into highly organized nanostructures. These structures are composed of a regular array of inorganic nanoparticles, usually with a 1- or 2-dimensional morphology, and held together by the organic species. It is intriguing to see that these structures on the one hand are characterized by a particularly tiny size of their inorganic building blocks, often in the size range of the unit cell, and on the other hand by a high order of the inorganic and organic component over a length scale of tens of nanometers up to several microns [168]. Selected examples are presented in Figure 3.10. The decomposition of a europium-benzoylacetonate complex in a mixture of oleic acid and oleylamine resulted in the formation of Eu_2O_3 nanoplatelets with a diameter of 32 nm and a thickness of 1.6 nm [122]. Redispersed in cyclohexane these platelets arranged into arrays with an interparticle distance of 2.3 nm (Figure 3.10a). Zinc oxide nanorods, obtained from zinc acetate, trioctylamine and oleic acid, self-organized on the TEM grid into isolated close-packed stacks with their long axis parallel to each other [157]. The rods had an average diameter of 2.2 nm and the length ranged from 40 to 50 nm (Figure 3.10b). Similarly gadolinium oxide nanoplatelets formed stacks on the TEM grid (Figure 3.10c). The particles with an edge length of 8 nm and a thickness of 1.1 nm were synthesized from gadolinium acetate in oleylamine, oleic acid and octadecene [12]. Samaria nanowires just 1.1 nm in diameter, however more than 1 μm long, aligned into parallel arrays (Figure 3.10d) [162].

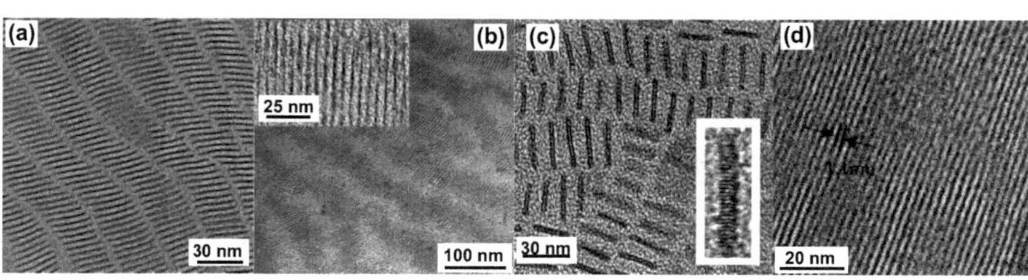

Fig. 3.10. **a–d** TEM images of various nanoparticle assemblies. **a** Arrays of Eu_2O_3 nanoplatelets. Image taken from Ref. [122] with permission of Wiley-VCH. **b** Arrays of 2 nm diameter ZnO nanorods (inset: higher magnification showing the oriented stacking). Image taken from Ref. [157]. **c** Stacks of Gd_2O_3 nanoplatelets (inset: HRTEM image). Image taken from Ref. [12]. **d** Arrays of rectangular-shaped samaria nanowires with a single unit cell thickness (1.1 nm). Image taken from Ref. [162]. Images **b–c** reproduced with permission of the American Chemical Society

3.8 Heterostructures/Multicomponent Nanoparticles

The main motivation behind nanoparticle research is the quest for new, or at least greatly improved, chemical and physical properties. With the demand for extending the functionality of nanoparticles, also the level of their structural and compositional complexity increases. The preparation of so-called hybrid nanocrystals or heteronanostructures, composed of two or more compositionally different materials sections and interconnected through permanent inorganic interfaces [13], constitutes a recent development in this direction. These heteronanostructures open up fascinating scientific and technological perspectives. The combination of subunits with varying properties (e.g., optical, magnetic, catalytic, or electronic) within one nanoparticle provides a "smart" platform, paving the way to the development of nanosized objects able to perform multiple tasks [27]. The electronic contacts between the different materials sections might lead to modified, or unexpected physical-chemical responses or coupling mechanisms, offering an additional tool to engineer the performance of these nanomaterials [13]. Especially for optoelectronic and photovoltaic applications combined semiconductors with different band gaps and appropriately chosen band offsets could either trap electron-hole pairs to force recombination at specific sites, or separate electrons from holes [88]. In the field of catalysis, connecting metal sections to metal oxides might lead to high efficiencies in redox reactions [146].

Core-shell configurations, in which one inorganic nanomaterial is uniformly grown around a compositionally different nanocrystal core, represent a relatively simple and widely studied family of heteronanostructures. Prerequisites for the formation of such structures is, on the one hand, a reasonable match between the lattice constants of both components, and, on the other hand, a selective heterogeneous nucleation of the shell onto the preformed core without further, regionally separated, homogeneous nucleation of the second material. There are two main strategies to core-shell structures. One involves the slow addition of the molecular precursor for the shell structure to the cores at relatively low temperatures. The second approach is based on the chemically or thermally induced conversion of the outermost layer of the core into the shell material. The different possible pathways made it possible to combine metals, semiconductors, magnetic and oxide materials, leading to heterostructures that, however, do not necessarily exhibit epitaxy between the components [27]. Most of the core-shell structures were synthesized with the purpose to increase the robustness and fluorescence efficiency of a semiconductor core. However, the deposition of a surface coating, which is then suitable for the attachment of organic molecules [77] or the tuning of the magnetic properties also have attracted much attention due to potential applications in biomedicine and spintronics [167]. An example for the latter case is the synthesis of bimagnetic core-shell $FePt/Fe_3O_4$ nanoparticles [165]. In a first step, $Fe_{58}Pt_{42}$ seeds were produced from $Pt(acac)_2$ and $Fe(CO)_5$ in octyl ether and subsequently coated with Fe_3O_4 by heating $Fe(acac)_3$, 1,2-

hexadecanediol, oleic acid, and oleylamine in phenyl ether. The thickness of the magnetite shell can be tuned by adjusting the $Fe(acac)_3$-to-seeds ratio. Figure 3.11a shows a TEM image of the $Fe_{58}Pt_{42}/Fe_3O_4$ core-shell nanoparticles with a 4 nm core and a 2 nm shell. The magnetic properties were found to be dependent on the shell thickness due to the exchange coupling between core and shell [165]. The decomposition of $Fe(CO)_5$ in octadecene in the presence of oleylamine yielded 6 nm-sized iron nanoparticles. Controlled oxidation of the iron surface in air produced spherical core-shell nanoparticles with an iron core of 4 nm and a magnetite shell of 2.5 nm (Figure 3.11b) [104]. Interestingly, these as-synthesized nanoparticles are amorphous and annealing under Ar is required to induce crystallization. Fe/Fe_3O_4 core-shell nanoparticles, however with a cube-like morphology, were prepared in squalene as solvent and sodium oleate as additive (Figure 3.11c) [118]. The core-shell structure was confirmed by HRTEM investigations and fast Fourier transform (FFT) filtering analysis (Figure 3.11c, inset). Lee et al. reported the synthesis of Ni/NiO core-shell nanoparticles from $Ni(acac)_2$ in oleylamine, TOPO and TOP [77]. Also in this case, first metal nanoparticles are produced, followed by oxidation in air. The NiO coating provides high affinity for polyhistidine, whereas the Ni core offers superparamagnetic properties. These nanoparticles have been successfully tested for separating and purifying His-tagged proteins from a multicomponent solution [77].

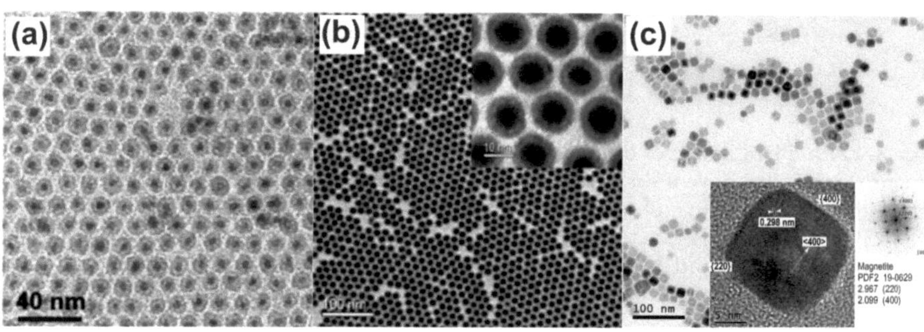

Fig. 3.11. **a** TEM image of $Fe_{58}Pt_{42}/Fe_3O_4$ core-shell nanoparticles with a 4 nm core and a 2 nm shell. Reproduced from Ref. [165] with permission of the American Chemical Society. **b** TEM image of Fe/Fe_3O_4 core-shell nanoparticles with a 4 nm core and 2.5 nm shell. Reproduced from Ref. [104] with permission of the American Chemical Society. **c** TEM overview image, HRTEM image and corresponding FFT pattern of Fe/Fe_3O_4 core-shell nanocubes. Image reproduced from Ref. [118] with permission of Wiley-VCH

In contrast to core-shell nanomaterials, in which only the outer shell surface is chemically accessible, noncentrosymmetric heterostructures have the advantage that all the subunits provide accessible functional moieties, analogous to organic molecules with several functional groups. One family of asymmetric heteronanostructures is represented by heterodimers or hetero-

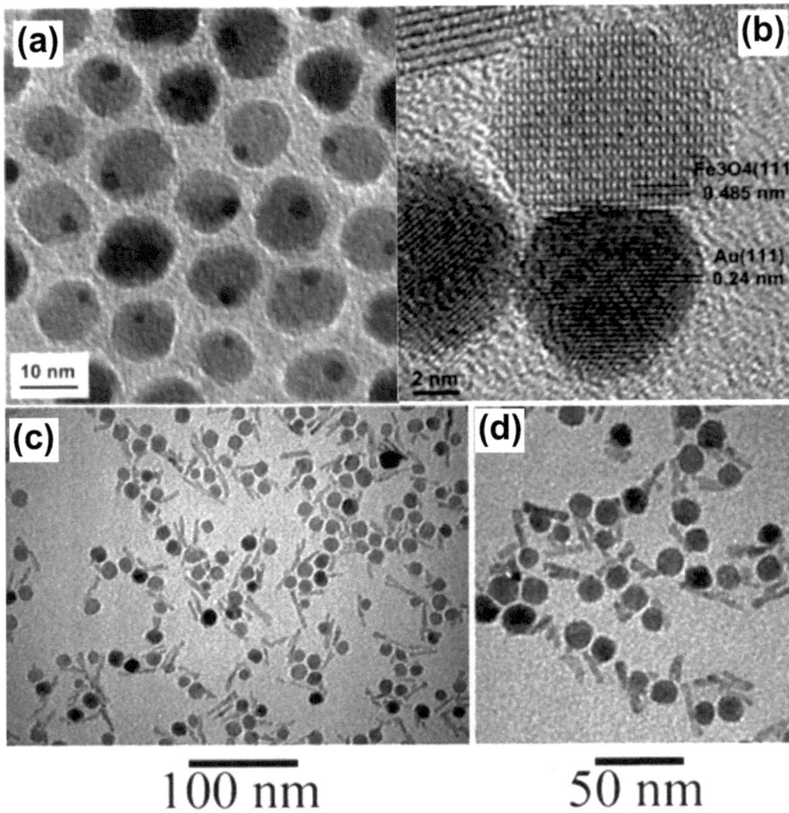

Fig. 3.12. **a** TEM image of dumbbell-like Au-Fe_3O_4 nanoparticles. The gold particles appear darker and have a diameter of 3 nm, whereas the iron oxide has a diameter of 14 nm. **b** HRTEM image of a 8–12 nm Au-Fe_3O_4 nanoparticle. **c** and **d** TEM images of γ-Fe_2O_3-TiO_2 heterostructures consisting of spherical iron oxide attached to titania nanorods. Images **a** and **b** reproduced from Ref. [159], **c** and **d** from Ref. [8] with permission of the American Chemical Society

oligomers, consisting of two or more nearly spherical nanocrystals, as reported for example for dumbbell-like bifunctional Au-Fe_3O_4 nanoparticles (Figure 3.12a) [159]. These bifunctional nanostructures were obtained by epitaxial growth of iron oxide on the gold seeds via decomposition of iron pentacarbonyl and oxidation in air. The final shape of the composite particles can be changed from dumbbell- to flower-like morphologies by varying the polarity of the solvent. Moreover, also the size of the individual spherical subunits can be tuned in a way that a rich variety of optical, magnetic, and chemical properties were achieved. Figure 3.12b shows a typical HRTEM image of one composite nanoparticle with a 12 nm Fe_3O_4 and a 8 nm Au subunit. Both particles are single-crystalline. However, it is also possible to combine units with different shapes. Semiconductor-magnetic heterostructures based on rod-like anatase TiO_2 sections with one γ-Fe_2O_3 sphere attached (Figure 3.12d) were prepared by the decomposition of iron pentacarbonyl in octadecene containing titania nanorods (obtained by an aqueous process) and the three surfactants oleic acid, oleylamine and 1,2-hexadecanediol in defined relative ratio [8]. By adjusting the reaction parameters, the size of the γ-Fe_2O_3 spheres that are heterogeneously nucleated on the TiO_2 seeds can finely be tuned, and thus the magnetic properties can also be controlled. Interestingly, the iron oxide domains exclusively grow on the longitudinal sides of the titania nanorods (Figure 3.12c and d), resulting in a high lattice misfit. Further information and many more examples regarding hybrid nanocrystals or heterostructures can be found in several excellent reviews [13, 27, 167].

3.9 Nonaqueous Processes Using Traces of Water

Although this book is dedicated to nonaqueous sol-gel processes, we want to briefly discuss some examples that use traces of water to round off the picture. As mentioned before, major problems of sol-gel methods based on the hydrolysis and condensation of molecular precursors are, on the one hand, the control over the fast reaction rates, and, on the other hand, the different reactivities of metal alkoxides, which makes it difficult to control the composition and the homogeneity of complex multicomponent oxides. One possibility to decrease and to adjust the reactivity of the precursors is the use of organic additives like carboxylic acids, β-diketones or functional alcohols, which act as chelating ligands and modify the reactivity of the precursors [86, 109, 47, 48, 51, 110]. An alternative strategy involves either the slow dissolution of water from the gas-phase into the synthesis mixture or addition of only small amounts of water. Although hydrolysis and condensation by the moisture of air has often been applied with great success [68, 89, 62], here only three representative examples are discussed in more detail.

Dicyclohexylzinc(II) $[Zn(c\text{-}C_6H_{11})_2]$, dissolved in tetrahydrofuran in the presence of different amines with a long alkyl chain, decomposed into crystalline ZnO nanoparticles at room temperature upon evaporation of the solvent (Figure 3.13a) [89]. The authors found that particle size, shape, and homogeneity were strongly influenced by the nature of the surfactant, the relative concentration of the reagents, the solvent, the overall concentration of the reagents, the reaction time, the evaporation time, and the reaction/evaporation temperature. The reaction of $Ce(NO_3)_3$ with a water/toluene/oleic acid mixture yielded CeO_2 nanocubes with an average size of 4.4 nm (Figure 3.13b) [153]. A similar particle morphology was obtained for Corundum-type In_2O_3 (Figure 3.13c), synthesized upon hydrolysis of indium isopropoxide using oleic acid, oleylamine, and trioctylamine as stabilizers [76].

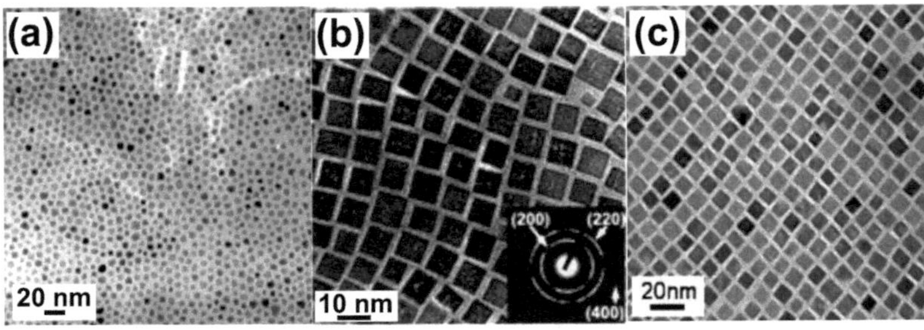

Fig. 3.13. a–c TEM overview images of metal oxide nanoparticles prepared in organic solvents in the presence of traces of water. **a** ZnO. Image taken from Ref. [89] with permission of Wiley-VCH. **b** CeO_2. Image taken from Ref. [153] with permission of the American Chemical Society. **c** In_2O_3. Image taken from Ref. [76] with permission of the American Chemical Society

3.10 Tabular Overview

The family of metal oxide nanoparticles synthesized via nonaqueous and surfactant-controlled processes has increased rapidly in the last few years, so that it is impossible to give a complete overview in this short book chapter. We tried to show, based on only a small number of selected and representative examples, the various synthesis methods within this active research field as well as their versatility to access complex nanomaterials with tailor-made properties. Table 3.1 gives a relatively exhaustive overview of metal oxide nanoparticles synthesized in organic solvents in the absence of water, together with precursors, solvents/surfactants, reaction conditions and morphological characteristics. Only liquid phase processes are listed, i.e., thermal decomposition reactions in air or in other gases are not considered. Some of the precursors are hydrated, which means that the water molecules may be responsible for the hydrolysis and condensation, rendering these approaches hydrolytic. Nevertheless, they were included in the table under the condition that no additional water was used in the reaction batch.

Table 3.1. Overview of metal oxide nanoparticles synthesized via surfactant-controlled nonaqueous processes

Metal oxide	Precursors	Solvents and Surfactants	T (°C)	Product Morphology	Ref.
Au/Fe_3O_4	$AuCl_3$, $Fe(CO)_5$	OA, OLA, ODA	180	Core/Hollow-shell nanoparticles	[120]
$BaTiO_3$	Metallic Ba, $Ti(O\textit{i}Pr)_4$	OA, DA or OLA	320	Spherical	[17]
CeO_2	$Ce(NO_3)_3{\cdot}6H_2O$	Ethanol, alkylamines, PVP	180	Spherical	[123]
CeO_2	$Ce(NO_3)_3{\cdot}6H_2O$	OA, TOA, diphenyl ether	320	Spherical and tadpole-shaped	[161]
CeO_2	Cerium-oleate complex	Various high-boiling solvents, OLA	180–290	Spherical	[37]
CeO_2	$(NH_4)_2Ce(NO_3)_6$	OLA, OA	230–300	Nanoflowers	[180]
CeO_2	Cerium nitrate hexahydrate	Octadecylamine	200–250	Nanoflowers or nanocubes	[143]
CeO_2	Cerium oleate	1-Hexadecene, OLA, octadecylamine	200	Nanorods or nanodumbbells	[1]
CeO_2	Cerium ammonium nitrate	OLA, OA,	230–300	Nanoflowers	[180]
CeO_2	Cerium nitrate hexahydrate	Methanol, decanoic acid	400	Spherical	[66]
CoO	$Co(acac)_3$	OA	200	Nanorods and pyramid-shaped	[117]
CoO	$Co_2(CO)_8$	THF, TEMPO and OLA	Room Temperature	Spherical	[74]
CoO	Cobalt oleate	OD	320	Pencil-shaped nanorods	[3]
CoO	Co_2O_3	OD, OLA, dodecanol	315	Tetrapods	[172]
CoO	Cobalt nitrate hexahydrate	Octadecylamine	250	Nanopolyhedrons	[143]
Co_3O_4	Cobalt nitrate hexahydrate	Octadecylamine	200	Nanocubes	[143]
CoO	$Co_2(CO)_8$	OLA, TOPO, o-dichlorobenzene, N-methylmorpholine	50	Spherical	[73]
Co_3O_4	$Co(NO_3)_2{\cdot}4H_2O$	n-Hexanol, SDBS, n-octanol	150–180	Spherical	[41, 42]
$CoFe_2O_4$	$Fe(acac)_3$, Fe(III) alkoxides, Co(II) isopropoxide	OLA, OA, benzyl ether	230–300	Spherical	[139]
$CoFe_2O_4$	$Fe(acac)_3$, $Co(acac)_{2,3}$	OLA, OA, 1,2-HD, tetraethylene glycol, benzyl ether	300	Spherical	[119]
$CoMn_2O_4$	$Co(acac)_2$, $Mn(acac)_3$	OA	230	Spherical	[169]
Co/MFe_2O_4 (M=Fe, Mn)	$Co_2(CO)_8$, $Fe(acac)_3$, $Mn(acac)_2$	1,2-HD, OLA, OA, benzyl ether	300	Core-shell	[105]
Cu_2O	$Cu(acac)_2$	OA	230	Spherical	[125]
Cu_2O	Copper acetate	OLA, TOA	270	Spherical	[158]
γ-Fe_2O_3, MnO	$Fe(Cup)_3$, $Mn(Cup)_2$	OA	300	Spherical	[108]
γ-Fe_2O_3	$Fe(CO)_5$	OLA/LA, octyl ether	100–120	Spherical	[50, 102]
γ-Fe_2O_3	$Fe(CO)_5$	OD, OLA, OA, 1,2-HD	240	Tetrapods	[28]
γ-Fe_2O_3	$Fe(NO_3)_3{\cdot}9H_2O$	LA	Melt	Nanorods	[56]
γ-Fe_2O_3	$Fe(CO)_5$	o-Dichlorobenzene, dodecylamine	–	Diverse well-defined shapes	[18]

Table 3.1. (continued)

Metal oxide	Precursors	Solvents and Surfactants	T (°C)	Product Morphology	Ref.
Fe_3O_4	$Fe(acac)_3$	OLA, OA, 1,2-HD, phenyl ether	Reflux	Spherical	[131, 176]
Fe_3O_4, Cr_2O_3, MnO, Co_3O_4, NiO	Metal fatty acid salts	OLA, OD	300	Various	[52]
Fe_3O_4	FeO(OH)	OLA, OD	320	Spherical	[163]
Fe_3O_4	$Fe(acac)_3$	Phenyl ether, octadecanol, OLA, OA	220	Faceted nanocrystals	[126]
FeO and Fe_3O_4	Iron acetate, $Fe(CO)_5$, $Fe(acac)_2$ or $Fe(acac)_3$	TOA, dioctyl ether, OLA	120–296	Spherical, faceted, cube-like, star-shaped	[107]
Fe_3O_4, MnO, CoO, FeO	Metal oleates	OLA, OD	320	Different shapes, but uniform within one system	[98, 72]
Fe_3O_4, MnO, CoO, NiO	$M(acac)_{2,3}$	Hexadecylamine, TOPO	220–270	Spherical (trigonal: NiO)	[81]
Fe_3O_4 and other iron oxide phases	$Fe(CO)_5$	Tridecanoic acid, octyl ether, 3-chloro peroxy-benzoic acid	293	Nanodiscs	[14]
Fe_3O_4	$Fe(acac)_3$	1,2-HD, OA, OLA, 1-adamantane-carboxylic acid, benzyl ether	260	Star- and flowerlike	[171]
Fe_3O_4	$Fe(CO)_5$	OD, OA	180	Hollow spheres	[103]
Fe_3O_4	Iron oleate	Sodium oleate, potassium oleate, dibutylammonium oleate, OLA, OD, diphenyl ether, tetracosane	290–375	Nanospheres or nanocubes	[70]
Fe_3O_4	$Fe(OC_2H_4OCH_3)_3$	OD, OLA	320	Spherical	[2]
Fe_3O_4	$Fe(acac)_3$	1,2-HD, OLA, OA, phenyl ether	254	Spherical	[43]
Fe_3O_4	$Fe(acac)_3$	OLA, OA, N-methylpyrrole	185	Spherical	[134]
$Fe_xFe_3O_4$, $Fe_{1-y}Mn_yO$-$Fe_{3-z}Mn_zO_4$	Iron(III) and manganese(II) formate hydrates	TOA, OLA	370	Nanocubes, nanostars, and other well-defined shapes	[44]
FePt/$CoFe_2O_4$	$Fe(acac)_3$, $Co(acac)_2$, $Pt(acac)_2$, $Fe(CO)_5$	Phenyl ether, 1,2-HD, OLA, OA	265	Bimagnetic nanobricks	[16]
Gd_2O_3	Gadolinium acetate	OA, OLA, OD	320	Nanoplates	[12]
Gd_2O_3	$Gd(acac)_3$	Palmitic acid, hexadecylamine, hydrazine monohydrate, TOA	320	Nanorings	[95]
Eu/Tb-doped Gd_2O_3	Metal acetate hydrates or acetylacetonate hydrates	OLA, OA, benzyl ether, 1,2-HD, TOPO	290	Nanoplates or nanospheres	[113]

Table 3.1. (continued)

Metal oxide	Precursors	Solvents and Surfactants	T (°C)	Product Morphology	Ref.
0HfO_2, ZrO_2, $Hf_xZr_{1-x}O_2$	Hf/Zr(O*i*Pr)$_4$·HO*i*Pr & Hf/$ZrCl_4$	TOPO	360	Spheres and rods	[135, 137, 138]
In_2O_3	In(acac)$_3$	OA	250	Spherical	[116]
In_2O_3	Indium acetate	OA, hexadecane, trimethylamine N-oxide	290	Spherical	[85]
In_2O_3, ZnO, CoO, MnO	Metal salt of a long-chain fatty acid	Octadecyl alcohol, OD	250–290	Nanoflowers	[91, 92]
In_2O_3	In(dipy)$_3$Cl$_3$·2H_2O	OA, OLA	290	Spherical	[181]
Indium tin oxide (ITO)	In(acac)$_3$, Sn(acac)$_2$Cl$_2$	OA	250	Spherical	[21]
Indium tin oxide	In(ac)$_3$, Sn(ac)$_2$	OD, myristic acid, octadecylamine	295	Spherical	[35]
$LiNbO_3$	LiNb(O*i*Pr)$_6$	TOPO	360	Nanorods	[147]
$LnFeO_3$ (Ln=La, Pr, Nd, Sm, Eu, Gd)	Ln-oleate, Fe-oleate	OD	320	20-100 nm	[150]
MFe_2O_4 (M = Fe, Co, Mn)	Fe(acac)$_3$, M(acac)$_2$	1,2-HD, OLA, OA, benzyl ether/octyl ether	Ca. 300	Spherical, cube- and polyhedron-shaped	[166, 128, 132, 142, 7, 154]
MFe_2O_4 (M = Co, Ni)	$FeCl_3$, Ni(ac)$_2$·4H_2O, Co(NO_3)$_2$·6H_2O	Sodium dodecylbenzene-sulfonate, 1-octanol	240	Spherical	[54]
$MnFe_2O_4$	$Mn_2(CO)_{10}$, $Fe(CO)_5$	OLA, octyl ether	300	Spherical	[63]
$MnFe_2O_4$	Mn(II) benzoylacetonate, Fe(III) benzoylacetonate	Phenyl ether, 1,2-HD, OLA, OA	260	Spherical	[127]
MnO	$Mn_2(CO)_{10}$	OA, TOP	300	Sperical and nanorods	[101]
MnO	Mn(oleate)$_2$	OLA, OA	320	Multipods	[182]
MnO, NiO	Mn(Cup)$_2$, Ni(Cup)$_2$	Toluene, TOPO	325 (MnO), 240 (NiO)	Spherical	[34]
MnO	Manganese formate hydrate	TOA, OLA	320	Dumbbell-shaped	[178]
Mn_3O_4, MnO	Mn(acac)$_2$	OA	180	Spherical	[115, 55]
Mn_3O_4, CoO, CuO_x	Metal formates M(HCOO)$_2$	OLA, OA	190	Spherical	[133]

Table 3.1. (continued)

Metal oxide	Precursors	Solvents and Surfactants	T (°C)	Product Morphology	Ref.
$NiFe_2O_4$	$Fe(acac)_3$, $Ni(acac)_2$, or iron(III) nitrate nonahydrate, nickel(II) nitrate hexahydrate	1,2-HD, OLA, OA, benzyl ether, or OA, 1,2-HD, OLA, phenyl ether	300	Spherical	[5]
$NiFe_2O_4$	$Ni(ac)_2{\cdot}4H_2O$, $Fe(ac)_3$	OLA, diphenyl ether, 1,12-dodecanediol, TOP	250	Spherical	[156]
NiO	$Ni(acac)_2$	OA, various tri-alkylphosphines	200	Spherical	[100]
NiO	Nickel nitrate hexahydrate	Octadecylamine	180	Nanoflowers	[143]
RE_2O_3 (RE = La to Lu, Y)	$RE(acac)_3$ (and others)	OLA, OD, OA	310–330	Nanoplates	[124, 122]
RE_2O_3 (RE = Pr, Nd, Sm, Eu, Gd, Tb, Dy)	Metal acetate or acetylacetonate	OA	Microwave	Nanoplates	[96]
RE_2O_3 (RE = Y, Dy, Ho, Er)	Rare earth nitrate hydrates	Dodecylamine, OD	320	Nanobelts	[38]
Sm_2O_3	Samarium acetate	OA, long-chain carboxylic acids	240	Assembled nanowires and nanoplates	[162]
SnO_2, ZnO	Tin or zinc 2-ethylhexanoate	Diphenyl ether & various amines	230–250	Spherical	[31]
SnO_2, TiO_2, In_2O_3, ZnO	Metal oxide sols	Tetradecene, dodecylamine	160	Spherical	[32]
SnO_2	$SnCl_4{\cdot}5H_2O$	OLA, OA, methanol or ethanol	180	Dots and nanowires	[151]
TiO_2	$TiCl_4$ & $Ti(OiPr)_4$	TOPO, heptadecane	300	Spherical	[140]
TiO_2	$Ti(OiPr)_3(dmae)$	TOPO	325	Spherical	[97]
TiO_2	$Ti(OiPr)_4$	OLA, toluene	250	Spheres to rods	[65]
TiO_2	$Ti(OiPr)_4$	OLA, ethyleneglycol	100	Spherical	[25]
TiO_2	$TiCl_4$ & $Ti(OiPr)_4$	Dioctyl ether, TOPO, LA	300	Bullet-shaped to nanorods	[60]
TiO_2	$Ti(COT)_2$	DMSO, TBP or TBPO or TOPO	120	Spherical	[136]
TiO_2	$Ti(OiPr)_4$	OLA, OD, OA	260	Nanorods	[175]
TiO_2	$Ti(OiPr)_4$	OLA	270	Nanorods	[58]
TiO_2	$Ti(OBu)_4$	NH_4HCO_3, LA, triethylamine, cyclohexane	150	Nanospheres and nanorods	[80]
TiO_2	$TiCl_4$, $Ti(OiPr)_4$	OD, OA	300	Nanorods	[69]
TiO_2- γ-Fe_2O_3	TiO_2 nanorods, $Fe(CO)_5$	OLA, OA, 1,2-dodecanediol	240–300	Heterostructures	[8]
TiO_2, ZrO_2	$M(OiPr)_4$ & MCl_4	TOPO, heptadecane	270	Spherical	[39]
TiO_2	$Ti(OiPr)_4$	Benzyl alcohol, various primary amines	180	Nanosheets	[148]

Table 3.1. (continued)

Metal oxide	Precursors	Solvents and Surfactants	T (°C)	Product Morphology	Ref.
TiO_2 (brookite)	$TiCl_4$	OD, OA, OLA	290	Nanorods	[9]
Zr-doped TiO_2	$TiCl_4$, $Ti(OiPr)_4$, $ZrCl_4$, $Zr(OiPr)_4$	TOPO	320–400	Nanorods	[15]
$W_{18}O_{49}$	$W(CO)_6$	OA	270	Nanorods	[78]
$W_{18}O_{49}$	$W(CO)_6$	OLA, hexadecylamine, TOPO, trimethylamine N-oxide	100	Nanorods	[145]
$W_{18}O_{49}$, TiO_2, Mn_3O_4, V_2O_5	Metal chlorides	OA, OLA,	300–350	Nanorods	[112, 111]
Y_2O_3:Eu	Metal carbonates	OLA, OA, TOPO	280	Spheres and discs	[144]
ZnO	Diethyl zinc	Decane, octylamine, TOPO	200	Spherical	[121]
ZnO	Zinc acetate	Various alkylamines, TBPA	220–300	Spherical	[24]
ZnO	Zinc acetate	Trioctylamine, OLA	286	Assembled nanorods	[157]
ZnO	ZnX_2 (X = Cl, Br)	OLA, hexadecylamine, dioctyl ether	180–250	Spherical	[26]
ZnO	ZnO	OD, OLA, hexadecylamine	310	Teardrops	[179]
ZnO	Zinc acetate	TOPO, dioctyl ether, 1,12-dodecanediol, hexadecylamine	250	Cone, hexagonal cone, rod-like	[57]
ZnO	Zinc acetate	OLA, trioctylamine, hexadecanol, octadecene	286–300	Spherical and nanorods	[4]
ZnO	Zn-oleate	OLA, OA	300	Hexagonal pyramid-shaped	[20]
ZnO	Zinc acetate	OA or hexadecylamine	240	Rods, wires or prisms	[174]
ZnO	Zinc acetate	Various	Various	Nanorods, nanotetrahedrons, nanofans, nanodumbbells, nanosquamas	[173]
ZnO	$Zn(acac)_2$	OA	205	"Heart"-shaped	[84]
Co-doped ZnO	Zinc acetate, cobalt acetate	TOA	310	Nanowires	[164, 22]
ZnO	Zinc acetate	OLA, oleyl alcohol	325	Tetrapod and spiked-clusters	[177]
ZnO	Zinc acetate dihydrate	1,2-Dodecanediol, benzyl ether, TOPO	280	Trigonal pyramids	[61]
ZnO	Zinc nitrate hexahydrate	Octadecylamine	200	Triangular nanoplates	[143]
ZnO	Zinc(II) oleate	OD, OLA	317	Nanopellets	[19]
ZnO	Zinc acetate	Dioctyl ether, TOPO, tetradecylphosphonic acid, 1,12-dodecanediol	250	Nanorods	[67]

Table 3.1. (continued)

Metal oxide	Precursors	Solvents and Surfactants	T (°C)	Product Morphology	Ref.
Eu-doped $ZnGa_2O_4$	Metal acetate hydrates, acetylacetonate hydrates, or metal nitrates	OLA, OA, benzyl ether, 1,2-HD, TOPO	290	Sperical, triangular, or rectangular	[114]
Mn-doped ZnO	Zinc acetate, $Mn(oleate)_2$	Benzyl alcohol, anisole	250	Spherical	[23]
Y_2O_3-stabilized ZrO_2	$Zr(OiPr)_4 \cdot HOiPr$, $ZrCl_4$, $Y(OiPr)_3$	OA	340	–	[36]
ZrO_2	$Zr(OiPr)_4 \cdot HOiPr$ & $ZrCl_4$	TOPO,	340	Spherical	[59]
$Zr_{1-x}Ce_xO_2$	$Zr(acac)_4$, $Ce(acac)_3$	OA	260	Spherical	[170]

Abbreviations: DA: Decanoic acid; OD: octadecene; THF: tetrahydrofuran; 1,2-HD: 1,2-hexadecanediol; TEMPO: 2,2,6,6-tetramethylpiperidine-1-oxyl; TOPO: trioctylphosphine oxide; TOP: trioctylphosphine; PVP: poly(vinylpyrrolidone); LA: lauric acid; COT: cyclooctatetraene; DMSO: dimethyl sulfoxide; TBP: tributylphosphine; TBPO: tributylphosphine oxide; TBPA: tert-butylphosphonic acid; OLA: oleic acid; OA: oleylamine; acac: acetylacetonate; Cup: N-nitrosophenylhydroxylamine; TOA: trioctylamine; dmae: dimethylaminoethoxide; SDBS: sodium dodecyl benzenesulfonate.

References

1. Ahniyaz, A., Sakamoto, Y., Bergström, L.: Tuning the aspect ratio of ceria nanorods and nanodumbbells by a face-specific growth and dissolution process. Cryst. Growth Des. **8**, 1798–1800 (2008)
2. Ahniyaz, A., Seisenbaeva, G.A., Häggström, L., Kamali, S., Kessler, V.G., Nordblad, P., Johansson, C., Bergström, L.: Preparation of iron oxide nanocrystals by surfactant-free or oleic acid-assisted thermal decomposition of a Fe(III) alkoxide. J. Magn. Magn. Mater. **320**, 781–787 (2008)
3. An, K., Lee, N., Park, J., Kim, S.C., Hwang, Y., Park, J.G., Kim, J.Y., Park, J.H., Han, M.J., Yu, J.J., Hyeon, T.: Synthesis, characterization, and self-assembly of pencil-shaped CoO nanorods. J. Am. Chem. Soc. **128**, 9753–9760 (2006)
4. Andelman, T., Gong, Y., Polking, M., Yin, M., Kuskovsky, I., Neumark, G., O'Brien, S.: Morphological control and photoluminescence of zinc oxide nanocrystals. J. Phys. Chem. B **109**, 14,314–14,318 (2005)
5. Baruwati, B., Manorama, S.V.: Monodispersed $NiFe_2O_4$ nanoparticles: Non-aqueous synthesis and characterization. Mater. Chem. Phys. **112**, 631–636 (2008)
6. Bentzon, M.D., Vonwonterghem, J., Tholen, A.: A magnetic super lattice. Ultramicroscopy **24**, 67–68 (1988)
7. Bhattacharyya, S., Salvetat, J.P., Fleurier, R., Husmann, A., Cacciaguerra, T., Saboungi, M.L.: One step synthesis of highly crystalline and high coercive cobalt-ferrite nanocrystals. Chem. Commun. pp. 4818–4820 (2005)
8. Buonsanti, R., Grillo, V., Carlino, E., Giannini, C., Curri, M.L., Innocenti, C., Sangregorio, C., Achterhold, K., Parak, F.G., Agostiano, A., Cozzoli, P.D.: Seeded growth of asymmetric binary nanocrystals made of a semiconductor TiO_2 rodlike section and a magnetic -Fe_2O_3 spherical domain. J. Am. Chem. Soc. **128**, 16,953–16,970 (2006)
9. Buonsanti, R., Grillo, V., Carlino, E., Giannini, C., Kipp, T., Cingolani, R., Cozzoli, P.D.: Nonhydrolytic synthesis of high-quality anisotropically shaped brookite TiO_2 nanocrystals. J. Am. Chem. Soc. **130**, 11,223–11,233 (2008)
10. Burda, C., Chen, X., Narayanan, R., El-Sayed, M.A.: Chemistry and properties of nanocrystals of different shapes. Chem. Rev. **105**, 1025–1102 (2005)
11. Byrappa, K., Adschiri, T.: Hydrothermal technology for nanotechnology. Prog. Cryst. Growth Charact. Mater. **53**, 117–166 (2007)
12. Cao, Y.C.: Synthesis of square gadolinium-oxide nanoplates. J. Am. Chem. Soc. **126**, 7456–7457 (2004)
13. Casavola, M., Buonsanti, R., Caputo, G., Cozzoli, P.D.: Colloidal strategies for preparing oxide-based hybrid nanocrystals. Eur. J. Inorg. Chem. pp. 837–854 (2008)
14. Casula, M.F., Jun, Y.W., Zaziski, D.J., Chan, E.M., Corrias, A., Alivisatos, A.P.: The concept of delayed nucleation in nanocrystal growth demonstrated for the case of iron oxide nanodisks. J. Am. Chem. Soc. **128**, 1675–1682 (2006)
15. Chang, S.M., Doong, R.A.: Characterization of Zr-doped TiO_2 nanocrystals prepared by a nonhydrolytic sol-gel method at high temperatures. J. Phys. Chem. B **110**, 20,808–20,814 (2006)
16. Chaubey, G.S., Nandwana, V., Poudyal, N., Rong, C.b., Liu, J.P.: Synthesis and characterization of bimagnetic bricklike nanoparticles. Chem. Mater. **20**, 475–478 (2008)
17. Chen, Z., Huang, L., He, J., Zhu, Y., O'Brien, S.: New nonhydrolytic route to synthesize crystalline $BaTiO_3$ nanocrystals with surface capping ligands. J. Mater. Res. **21**, 3187–3195 (2006)

18. Cheon, J., Kang, N.J., Lee, S.M., Lee, J.H., Yoon, J.H., Oh, S.J.: Shape evolution of single-crystalline iron oxide nanocrystals. J. Am. Chem. Soc. **126**, 1950–1951 (2004)
19. Chiu, W., Khiew, P., Isa, D., Cloke, M., Radiman, S., Abd-Shukor, R., Abdullah, M., Huang, N.: Synthesis of two-dimensional ZnO nanopellets by pyrolysis of zinc oleate. Chem. Eng. J. **142**, 337–343 (2008)
20. Choi, S.H., Kim, E.G., Park, J., An, K., Lee, N., Kim, S.C., Hyeon, T.: Large-scale synthesis of hexagonal pyramid-shaped ZnO nanocrystals from thermolysis of Zn-oleate complex. J. Phys. Chem. B **109**, 14,792–14,794 (2005)
21. Choi, S.I., Nam, K.M., Park, B.K., Seo, W.S., Park, J.T.: Preparation and optical properties of colloidal, monodisperse, and highly crystalline ITO nanoparticles. Chem. Mater. **20**, 2609–2611 (2008)
22. Clavel, G., Pinna, N., Zitoun, D.: Magnetic properties of cobalt and manganese doped ZnO nanowires. Phys. Stat. Sol. A **204**, 118–124 (2007)
23. Clavel, G., Willinger, M.G., Zitoun, D., Pinna, N.: Solvent dependent shape and magnetic properties of doped ZnO nanostructures. Adv. Funct. Mater. **17**, 3159–3169 (2007)
24. Cozzoli, P.D., Curri, M.L., Agostiano, A., Leo, G., Lomascolo, M.: ZnO nanocrystals by a non-hydrolytic route: Synthesis and characterization. J. Phys. Chem. B **107**, 4756–4762 (2003)
25. Cozzoli, P.D., Kornowski, A., Weller, H.: Low-temperature synthesis of soluble and processable organic-capped anatase TiO_2 nanorods. J. Am. Chem. Soc. **125**, 14,539–14,548 (2003)
26. Cozzoli, P.D., Kornowski, A., Weller, H.: Colloidal synthesis of organic-capped ZnO nanocrystals via a sequential reduction-oxidation reaction. J. Phys. Chem. B **109**, 2638–2644 (2005)
27. Cozzoli, P.D., Pellegrino, T., Manna, L.: Synthesis, properties and perspectives of hybrid nanocrystal structures. Chem. Soc. Rev. **35**, 1195–1208 (2006)
28. Cozzoli, P.D., Snoeck, E., Garcia, M.A., Giannini, C., Guagliardi, A., Cervellino, A., Gozzo, F., Hernando, A., Achterhold, K., Ciobanu, N., Parak, F.G., Cingolani, R., Manna, L.: Colloidal synthesis and characterization of tetrapod-shaped magnetic nanocrystals. Nano Lett. **6**, 1966–1972 (2006)
29. Demazeau, G.: Solvothermal reactions: An original route for the synthesis of novel materials. J. Mater. Sci. **43**, 2104–2114 (2008)
30. Ding, K.L., Miao, Z.J., Liu, Z.M., Zhang, Z.F., Han, B.X., An, G.M., Miao, S.D., Xie, Y.: Facile synthesis of high quality TiO_2 nanocrystals in ionic liquid via a microwave-assisted process. J. Am. Chem. Soc. **129**, 6362–6363 (2007)
31. Epifani, M., Arbiol, J., Díaz, R., Perálvarez, M.J., Siciliano, P., Morante, J.R.: Synthesis of SnO_2 and ZnO colloidal nanocrystals from the decomposition of tin(II) 2-ethylhexanoate and zinc(II) 2-ethylhexanoate. Chem. Mater. **17**, 6468–6472 (2005)
32. Epifani, M., Díaz, R., Arbiol, J., Comini, E., Sergent, N., Pagnier, T., Siciliano, P., Faglia, G., Morante, J.R.: Nanocrystalline metal oxides from the injection of metal oxide sols in coordinating solutions: Synthesis, characterization, thermal stabilization, device processing, and gas-sensing properties. Adv. Funct. Mater. **16**, 1488–1498 (2006)
33. Gerbec, J.A., Magana, D., Washington, A., Strouse, G.F.: Microwave-enhanced reaction rates for nanoparticle synthesis. J. Am. Chem. Soc. **127**, 15,791–15,800 (2005)
34. Ghosh, M., Biswas, K., Sundaresan, A., Rao, C.N.R.: MnO and NiO nanoparticles: Synthesis and magnetic properties. J. Mater. Chem. **16**, 106–111 (2006)
35. Gilstrap Jr., R.A., Capozzi, C.J., Carson, C.G., Gerhardt, R.A., Summers, C.J.: Synthesis of a nonagglomerated indium tin oxide nanoparticle dispersion. Adv. Mater. **20**, 4163–4166 (2008)

36. Goto, Y., Omata, T., Otsuka-Yao-Matsuo, S.: Extremely suppressed grain growth of Y_2O_3-stabilized zirconia nanocrystals synthesized by the nonhydrolytic sol-gel technique. J. Electrochem. Soc. **156**, K4–K9 (2009)
37. Gu, H., Soucek, M.D.: Preparation and characterization of monodisperse cerium oxide nanoparticles in hydrocarbon solvents. Chem. Mater. **19**, 1103–1110 (2007)
38. Han, M., Shi, N.E., Zhang, W.L., Li, B.J., Sun, J.H., Chen, K.J., Zhu, J.M., Wang, X., Xu, Z.: Large-scale synthesis of single-crystalline RE_2O_3 (RE=Y, Dy, Ho, Er) nanobelts by a solid-liquid-phase chemical route. Chem. Eur. J. **14**, 1615–1620 (2007)
39. Hao, E.C., Anderson, N.A., Asbury, J.B., Lian, T.Q.: Effect of trap states on interfacial electron transfer between molecular absorbates and semiconductor nanoparticles. J. Phys. Chem. B **106**, 10,191–10,198 (2002)
40. Hayes, B.L.: Microwave synthesis: chemistry at the speed of light. CEM Publishing: Matthews, USA (2002)
41. He, T., Chen, D., Jiao, X.: Controlled synthesis of Co_3O_4 nanoparticles through oriented aggregation. Chem. Mater. **16**, 737–743 (2004)
42. He, T., Chen, D., Jiao, X., Wang, Y., Duan, Y.: Solubility-controlled synthesis of high-quality Co_3O_4 nanocrystals. Chem. Mater. **17**, 4023–4030 (2005)
43. Herranz, F., Morales, M., Roca, A., Desco, M., Ruiz-Cabello, J.: A new method for the rapid synthesis of water stable superparamagnetic nanoparticles. Chem. Eur. J. **14**, 9126–9130 (2008)
44. Hofmann, C., Rusakova, I., Ould-Ely, T., Prieto-Centurión, D., Hartman, K.B., Kelly, A.T., Lüttge, A., Whitmire, K.H.: Shape control of new Fe_xO-Fe_3O_4 and $Fe_{1-y}Mn_yO$-$Fe_{3-z}Mn_zO_4$ nanostructures. Adv. Funct. Mater. **18**, 1661–1667 (2008)
45. Hou, Y., Gao, S., Ohta, T., Kondoh, H.: Towards 3-D spherical self-assembly by ternary surfactant combinations: The case of magnetite nanoparticles. Eur. J. Inorg. Chem. pp. 1169–1173 (2004)
46. Hou, Y.L., Yu, J.F., Gao, S.: Solvothermal reduction synthesis and characterization of superparamagnetic magnetite nanoparticles. J. Mater. Chem. **13**, 1983–1987 (2003)
47. Hubert-Pfalzgraf, L.G.: Some aspects of homo and heterometallic alkoxides based on functional alkohols. Coord. Chem. Rev. **178–180**, 967–997 (1998)
48. Hubert-Pfalzgraf, L.G.: To what extent can design of molecular precursors control the preparation of high tech oxides. J. Mater. Chem. **14**, 3113–3123 (2004)
49. Hyeon, T.: Chemical synthesis of magnetic nanoparticles. Chem. Commun. pp. 927–934 (2003)
50. Hyeon, T., Lee, S.S., Park, J., Chung, Y., Na, H.B.: Synthesis of highly crystalline and monodisperse maghemite nanocrystallites without size-selection process. J. Am. Chem. Soc. **123**, 12,798–12,801 (2001)
51. In, M., Sanchez, C.: Growth versus cyclization in the early stages of polycondensation of metal alkoxides. J. Phys. Chem. B **109**, 23,870–23,878 (2005)
52. Jana, N.R., Chen, Y., Peng, X.: Size- and shape-controlled magnetic (Cr, Mn, Fe, Co, Ni) oxide nanocrystals via a simple and general approach. Chem. Mater. **16**, 3931–3935 (2004)
53. Jana, N.R., Gearheart, L., Murphy, C.J.: Wet chemical synthesis of silver nanorods and nanowires of controllable aspect ratio. Chem. Commun. pp. 617–618 (2001)
54. Jia, X., Chen, D.R., Jiao, X.L., He, T., Wang, H.Y., Jiang, W.: Monodispersed Co, Ni-ferrite nanoparticles with tunable sizes: Controlled synthesis, magnetic properties, and surface modification. J. Phys. Chem. C **112**, 911–917 (2008)
55. Jiao, F., Harrison, A., Bruce, P.G.: Ordered three-dimensional arrays of monodispersed Mn_3O_4 nanoparticles with a core-shell structure and spin-glass behavior. Angew. Chem. Int. Ed. **46**, 3946–3950 (2007)

56. Jing, Z., Wu, S.: Synthesis, characterization and magnetic properties of γ-Fe_2O_3 nanoparticles via a non-aqueous medium. J. Solid State Chem. **177**, 1213–1218 (2004)
57. Joo, J., Kwon, S.G., Yu, J.H., Hyeon, T.: Synthesis of ZnO nanocrystals with cone, hexagonal cone, and rod shapes via non-hydrolytic ester elimination sol-gel reactions. Adv. Mater. **17**, 1873–1877 (2005)
58. Joo, J., Kwon, S.G., Yu, T., Cho, M., Lee, J., Yoon, J., Hyeon, T.: Large-scale synthesis of TiO_2 nanorods via nonhydrolytic sol-gel ester elimination reaction and their application to photocatalytic inactivation of E. coli. J. Phys. Chem. B **109**, 15,297–15,302 (2005)
59. Joo, J., Yu, T., Kim, Y.W., Park, H.M., Wu, F., Zhang, J.Z., Hyeon, T.: Multigram scale synthesis and characterization of monodisperse tetragonal zirconia nanocrystals. J. Am. Chem. Soc. **125**, 6553–6557 (2003)
60. Jun, Y.W., Casula, M.F., Sim, J.H., Kim, S.Y., Cheon, J., Alivisatos, A.P.: Surfactant-assisted elimination of a high energy facet as a means of controlling the shapes of TiO_2 nanocrystals. J. Am. Chem. Soc. **125**, 15,981–15,985 (2003)
61. Kachynski, A.V., Kuzmin, A.N., Nyk, M., Roy, I., Prasad, P.N.: Zinc oxide nanocrystals for nonresonant nonlinear optical microscopy in biology and medicine. J. Phys. Chem. C **112**, 10,721–10,724 (2008)
62. Kahn, M.L., Monge, M., Snoeck, E., Maisonnat, A., Chaudret, B.: Spontaneous formation of ordered 2D and 3D superlattices of ZnO nanocrystals. Small **1**, 221–224 (2005)
63. Kang, E., Park, J., Hwang, Y., Kang, M., Park, J.G., Hyeon, T.: Direct synthesis of highly crystalline and monodisperse manganese ferrite nanocrystals. J. Phys. Chem. B **108**, 13,932–13,935 (2004)
64. Kappe, C.O.: Controlled microwave heating in modern organic synthesis. Angew. Chem. Int. Ed. **43**, 6250–6284 (2004)
65. Kim, C.S., Moon, B.K., Park, J.H., Choi, B.C., Seo, H.J.: Solvothermal synthesis of nanocrystalline TiO_2 in toluene with surfactant. J. Cryst. Growth **257**, 309–315 (2003)
66. Kim, J., Park, Y.S., Veriansyah, B., Kim, J.D., Lee, Y.W.: Continuous synthesis of surface-modified metal oxide nanoparticles using supercritical methanol for highly stabilized nanofluids. Chem. Mater. **20**, 6301–6303 (2008)
67. Kim, S.Y., Yeon, Y.S., Park, S.M., Kim, J.H., Song, J.K.: Exciton states of quantum confined ZnO nanorods. Chem. Phys. Lett. **462**, 100–103 (2008)
68. Kominami, H., Kohno, M., Takada, Y., Inoue, M., Inui, T., Kera, Y.: Hydrolysis of titanium alkoxide in organic solvent at high temperatures: A new synthetic method for nanosized, thermally stable titanium(IV) oxide. Ind. Eng. Chem. Res. **38**, 3925–3931 (1999)
69. Koo, B., Park, J., Kim, Y., Choi, S.H., Sung, Y.E., Hyeon, T.: Simultaneous phase- and size-controlled synthesis of TiO_2 nanorods via non-hydrolytic sol-gel reaction of syringe pump delivered precursors. J. Phys. Chem. B **110**, 24,318–24,323 (2006)
70. Kovalenko, M.V., Bodnarchuk, M.I., Lechner, R.T., Hesser, G., Schäffler, F., Heiss, W.: Fatty acid salts as stabilizers in size- and shape-controlled nanocrystal synthesis: The case of inverse spinel iron oxide. J. Am. Chem. Soc. **129**, 6352–6353 (2007)
71. Kumar, S., Nann, T.: Shape control of II-VI semiconductor nanomaterials. Small **2**, 316–329 (2006)
72. Kwon, S.G., Piao, Y., Park, J., Angappane, S., Jo, Y., Hwang, N.M., Park, J.G., Hyeon, T.: Kinetics of monodisperse iron oxide nanocrystal formation by "heating-up" process. J. Am. Chem. Soc. **129**, 12,571–12,584 (2007)
73. Lagunas, A., Mairata i Payeras, A., Jimeno, C., Puntes, V.F., Pericàs, M.A.: Low-temperature synthesis of CoO nanoparticles via chemically assisted oxidative decarbonylation. Chem. Mater. **20**, 92–100 (2008)

74. Lagunas, A., Payeras, A.M.i., Jimeno, C., Pericas, M.A.: TEMPO-mediated, room temperature synthesis of pure CoO nanoparticles. Chem. Commun. pp. 1307–1309 (2006)
75. Latham, A.H., Williams, M.E.: Controlling transport and chemical functionality of magnetic nanoparticles. Acc. Chem. Res. **41**, 411–420 (2008)
76. Lee, C.H., Kim, M., Kim, T., Kim, A., Paek, J., Lee, J.W., Choi, S.Y., Kim, K., Park, J.B., Lee, K.: Ambient pressure syntheses of size-controlled corundum-type In_2O_3 nanocubes. J. Am. Chem. Soc. **128**, 9326–9327 (2006)
77. Lee, I.S., Lee, N., Park, J., Kim, B.H., Yi, Y.W., Kim, T., Kim, T.K., Lee, I.H., Paik, S.R., Hyeon, T.: Ni/NiO core/shell nanoparticles for selective binding and magnetic separation of histidine-tagged proteins. J. Am. Chem. Soc. **128**, 10,658–10,659 (2006)
78. Lee, K., Seo, W.S., Park, J.T.: Synthesis and optical properties of colloidal tungsten oxide nanorods. J. Am. Chem. Soc. **125**, 3408–3409 (2003)
79. Leonelli, C., Lojkowski, W.: Main development directions in the application of microwave irradiation to the synthesis of nanopowders. Chimica Oggi-Chemistry Today **25**, 34–38 (2007)
80. Li, X.L., Peng, Q., Yi, J.X., Wang, X., Li, Y.: Near monodisperse TiO_2 nanoparticles and nanorods. Chem. Eur. J. **12**, 2383–2391 (2006)
81. Li, Y., Afzaal, M., O'Brien, P.: The synthesis of amine-capped magnetic (Fe, Mn, Co, Ni) oxide nanocrystals and their surface modification for aqueous dispersibility. J. Mater. Chem. **16**, 2175–2180 (2006)
82. Lian, S.Y., Wang, E.B., Gao, L., Wu, D., Song, Y.L., Xu, L.: Surfactant-assisted solvothermal preparation of submicrometer-sized hollow hematite particles and their photocatalytic activity. Mater. Res. Bull. **41**, 1192–1198 (2006)
83. Lian, S.Y., Wang, E.B., Gao, L., Xu, L.: Fabrication of single-crystalline Co_3O_4 nanorods via a low-temperature solvothermal process. Mater. Lett. **61**, 3893–3896 (2007)
84. Liu, J.F., Bei, Y.Y., Wu, H.P., Shen, D., Gong, J.Z., Li, X.G., Wang, Y.W., Jiang, N.P., Jiang, J.Z.: Synthesis of relatively monodisperse ZnO nanocrystals from a precursor zinc 2,4-pentanedionate. Mater. Lett. **61**, 2837–2840 (2007)
85. Liu, Q., Lu, W., Ma, A., Tang, J., Lin, J., Fang, J.: Study of quasi-monodisperse In_2O_3 nanocrystals: synthesis and optical determination. J. Am. Chem. Soc. **127**, 5276–5277 (2005)
86. Livage, J., Henry, M., Sanchez, C.: Sol-gel chemistry of transition metal oxides. Prog. Solid State Chem. **18**, 259–341 (1988)
87. de Mello Donega, C., Liljeroth, P., Vanmaekelbergh, D.: Physicochemical evaluation of the hot-injection method, a synthesis route for monodisperse nanocrystals. Small **1**, 1152–1162 (2005)
88. Milliron, D.J., Hughes, S.M., Cui, Y., Manna, L., Li, J.B., Wang, L.W., Alivisatos, A.P.: Colloidal nanocrystal heterostructures with linear and branched topology. Nature **430**, 190–195 (2004)
89. Monge, M., Kahn, M.L., Maisonnat, A., Chaudret, B.: Room-temperature organometallic synthesis of soluble and crystalline ZnO nanoparticles of controlled size and shape. Angew. Chem. Int. Ed. **42**, 5321–5324 (2003)
90. Murray, C.B., Norris, D.J., Bawendi, M.G.: Synthesis and characterization of nearly monodisperse CdE (E=S, Se, Te) semiconductor nanocrystallites. J. Am. Chem. Soc. **115**, 8706–8715 (1993)
91. Narayanaswamy, A., Xu, H., Pradhan, N., Kim, M., Peng, X.: Formation of nearly monodisperse In_2O_3 nanodots and oriented-attached nanoflowers: Hydrolysis and alcoholysis vs pyrolysis. J. Am. Chem. Soc. **128**, 10,310–10,319 (2006)
92. Narayanaswamy, A., Xu, H., Pradhan, N., Peng, X.: Crystalline nanoflowers with different chemical compositions and physical properties grown by limited ligand protection. Angew. Chem. Int. Ed. **45**, 5361–5364 (2006)

93. Niederberger, M., Garnweitner, G., Pinna, N., Antonietti, M.: Nonaqueous and halide-free route to crystalline $BaTiO_3$, $SrTiO_3$, and $(Ba,Sr)TiO_3$ nanoparticles via a mechanism involving C-C bond formation. J. Am. Chem. Soc. **126**, 9120–9126 (2004)
94. Niederberger, M., Pinna, N., Polleux, J., Antonietti, M.: A general soft chemistry route to perovskites and related materials: Synthesis of $BaTiO_3$, $BaZrO_3$ and $LiNbO_3$ nanoparticles. Angew. Chem. Int. Ed. **43**, 2270–2273 (2004)
95. Paek, J., Lee, C.H., Choi, J., Choi, S.Y., Kim, A., Lee, J.W., Lee, K.: Gadolinium oxide nanoring and nanoplate: Anisotropic shape control. Cryst. Growth Des. **7**, 1378–1380 (2007)
96. Panda, A.B., Glaspell, G., El-Shall, M.S.: Microwave synthesis and optical properties of uniform nanorods and nanoplates of rare earth oxides. J. Phys. Chem. C **111**, 1861–1864 (2007)
97. Parala, H., Devi, A., Bhakta, R., Fischer, R.A.: Synthesis of nano-scale TiO_2 particles by a nonhydrolytic approach. J. Mater. Chem. **12**, 1625–1627 (2002)
98. Park, J., An, K., Hwang, Y., Park, J.G., Noh, H.J., Kim, J.Y., Park, J.H., Hwang, N.M., Hyeon, T.: Ultra-large-scale syntheses of monodisperse nanocrystals. Nature Mater. **3**, 891–895 (2004)
99. Park, J., Joo, J., Kwon, S.G., Jang, Y., Hyeon, T.: Synthesis of monodisperse spherical nanocrystals. Angew. Chem. Int. Ed. **46**, 4630–4660 (2007)
100. Park, J., Kang, E., Son, S.U., Park, H.M., Lee, M.K., Kim, J., Kim, K.W., Noh, H.J., Park, J.H., Bae, C.J., Park, J.G., Hyeon, T.: Monodisperse nanoparticles of Ni and NiO: Synthesis, characterization, self-assembled superlattices, and catalytic applications in the suzuki coupling reaction. Adv. Mater. **17**, 429–434 (2005)
101. Park, J., Kang, E.A., Bae, C.J., Park, J.G., Noh, H.J., Kim, J.Y., Park, J.H., Park, J.H., Hyeon, T.: Synthesis, characterization, and magnetic properties of uniform-sized MnO nanospheres and nanorods. J. Phys. Chem. B **108**, 13,594–13,598 (2004)
102. Park, J., Lee, E., Hwang, N.M., Kang, M., Kim, S.C., Hwang, Y., Park, J.G., Noh, H.J., Kim, J.Y., Park, J.H., Hyeon, T.: One-nanometer-scale size-controlled synthesis of monodisperse magnetic iron oxide nanoparticles. Angew. Chem. Int. Ed. **44**, 2872–2877 (2005)
103. Peng, S., Sun, S.H.: Synthesis and characterization of monodisperse hollow Fe_3O_4 nanoparticles. Angew. Chem. Int. Ed. **46**, 4155–4158 (2007)
104. Peng, S., Wang, C., Xie, J., Sun, S.H.: Synthesis and stabilization of monodisperse Fe nanoparticles. J. Am. Chem. Soc. **128**, 10,676–10,677 (2006)
105. Peng, S., Xie, J., Sun, S.: Synthesis of Co/MFe_2O_4 (M=Fe, Mn) core/shell nanocomposite particles. J. Solid State Chem. **181**, 1560–1564 (2008)
106. Rajamathi, M., Seshadri, R.: Oxide and chalcogenide nanoparticles from hydrothermal/solvothermal reactions. Curr. Opin. Solid State Mater. Sci. **6**, 337–345 (2002)
107. Redl, F.X., Black, C.T., Papaefthymiou, G.C., Sandstrom, R.L., Yin, M., Zeng, H., Murray, C.B., O'Brien, S.P.: Magnetic, electronic, and structural characterization of nonstoichiometric iron oxides at the nanoscale. J. Am. Chem. Soc. **126**, 14,583–14,599 (2004)
108. Rockenberger, J., Scher, E.C., Alivisatos, A.P.: A new nonhydrolytic single-precursor approach to surfactant-capped nanocrystals of transition metal oxides. J. Am. Chem. Soc. **121**, 11,596–11,597 (1999)
109. Sanchez, C., Livage, J., Henry, M., Babonneau, F.: Chemical modification of alkoxide precursors. J. Non-Cryst. Solids **100**, 65–76 (1988)
110. Schubert, U.: Chemical modification of titanium alkoxides for sol-gel processing. J. Mater. Chem. **15**, 3701–3715 (2005)

111. Seo, J.W., Chung, H., Kim, M.Y., Lee, J., Choi, I.H., Cheon, J.: Development of water-soluble single-crystalline TiO_2 nanoparticles for photocatalytic cancer-cell treatment. Small **3**, 850–853 (2007)
112. Seo, J.W., Jun, Y.W., Ko, S.J., Cheon, J.: In situ one-pot synthesis of 1-dimensional transition metal oxide nanocrystals. J. Phys. Chem. B **109**, 5389–5391 (2005)
113. Seo, S., Yang, H., Holloway, P.H.: Controlled shape growth of Eu- or Tb-doped luminescent Gd_2O_3 colloidal nanocrystals. J. Colloid Interface Sci. **331**, 236–242 (2009)
114. Seo, S., Yang, H., Holloway, P.H.: Synthesis and properties of colloidal ternary $ZnGa_2O_4$:Eu^{3+} nanocrystals. J. Lumin. **129**, 307–311 (2009)
115. Seo, W.S., Jo, H.H., Lee, K., Kim, B., Oh, S.J., Park, J.T.: Size-dependent magnetic properties of colloidal Mn_3O_4 and MnO nanoparticles. Angew. Chem. Int. Ed. **43**, 1115–1117 (2004)
116. Seo, W.S., Jo, H.H., Lee, K., Park, J.T.: Preparation and optical properties of highly crystalline, colloidal, and size-controlled indium oxide nanoparticles. Adv. Mater. **15**, 795–797 (2003)
117. Seo, W.S., Shim, J.H., Oh, S.J., Lee, E.K., Hur, N.H., Park, J.T.: Phase- and size-controlled synthesis of hexagonal and cubic CoO nanocrystals. J. Am. Chem. Soc. **127**, 6188–6189 (2005)
118. Shavel, A., Rodriguez-Gonzalez, B., Spasova, M., Farle, M., Liz-Marzan, L.M.: Synthesis and characterization of iron/iron oxide core/shell nanocubes. Adv. Funct. Mater. **17**, 3870–3876 (2007)
119. Shemer, G., Tirosh, E., Livneh, T., Markovich, G.: Tuning a colloidal synthesis to control Co^{2+} doping in ferrite nanocrystals. J. Phys. Chem. C **111**, 14,334–14,338 (2007)
120. Shevchenko, E.V., Bodnarchuk, M.I., Kovalenko, M.V., Talapin, D.V., Smith, R.K., Aloni, S., Heiss, W., Alivisatos, A.P.: Gold/iron oxide core/hollow-shell nanoparticles. Adv. Mater. **20**, 4323–4329 (2008)
121. Shim, M., Guyot-Sionnest, P.: Organic-capped ZnO nanocrystals: Synthesis and n-type character. J. Am. Chem. Soc. **123**, 11,651–11,654 (2001)
122. Si, R., Zhang, Y.W., You, L.P., Yan, C.H.: Rare-earth oxide nanopolyhedra, nanoplates, and nanodisks. Angew. Chem. Int. Ed. **44**, 3256–3260 (2005)
123. Si, R., Zhang, Y.W., You, L.P., Yan, C.H.: Self-organized monolayer of nanosized ceria colloids stabilized by poly(vinylpyrrolidone). J. Phys. Chem. B **110**, 5994–6000 (2006)
124. Si, R., Zhang, Y.W., Zhou, H.P., Sun, L.D., Yan, C.H.: Controlled-synthesis, self-assembly behavior, and surface-dependent optical properties of high-quality rare-earth oxide nanocrystals. Chem. Mater. **19**, 18–27 (2007)
125. Son, S.U., Park, I.K., Park, J., Hyeon, T.: Synthesis of Cu_2O coated Cu nanoparticles and their successful applications to Ullmann-type amination coupling reactions of aryl chlorides. Chem. Commun. pp. 778–779 (2004)
126. Song, Q., Ding, Y., Wang, Z.L., Zhang, Z.J.: Formation of orientation-ordered superlattices of magnetite magnetic nanocrystals from shape-segregated self-assemblies. J. Phys. Chem. B **110**, 25,547–25,550 (2006)
127. Song, Q., Ding, Y., Wang, Z.L., Zhang, Z.J.: Tuning the thermal stability of molecular precursors for the nonhydrolytic synthesis of magnetic $MnFe_2O_4$ spinel nanocrystals. Chem. Mater. **19**, 4633–4638 (2007)
128. Song, Q., Zhang, Z.J.: Shape control and associated magnetic properties of spinel cobalt ferrite nanocrystals. J. Am. Chem. Soc. **126**, 6164–6168 (2004)
129. Strauss, C.R., Varma, R.S.: Microwaves in green and sustainable chemistry. Top. Curr. Chem. **266**, 199–231 (2006)
130. Sun, C.W., Li, H., Zhang, H.R., Wang, Z.X., Chen, L.Q.: Controlled synthesis of CeO_2 nanorods by a solvothermal method. Nanotechnology **16**, 1454–1463 (2005)

131. Sun, S.H., Zeng, H.: Size-controlled synthesis of magnetite nanoparticles. J. Am. Chem. Soc. **124**, 8204–8205 (2002)
132. Sun, S.H., Zeng, H., Robinson, D.B., Raoux, S., Rice, P.M., Wang, S.X., Li, G.X.: Monodisperse MFe_2O_4 (M = Fe, Co, Mn) nanoparticles. J. Am. Chem. Soc. **126**, 273–279 (2004)
133. Sun, X., Zhang, Y.W., Si, R., Yan, C.H.: Metal (Mn, Co, and Cu) oxide nanocrystals from simple formate precursors. Small **1**, 1081–1086 (2005)
134. Tan, Y., Zhuang, Z., Peng, Q., Li, Y.: Room-temperature soft magnetic iron oxide nanocrystals: Synthesis, characterization, and size-dependent magnetic properties. Chem. Mater. **20**, 5029–5034 (2008)
135. Tang, J., Fabbri, J., Robinson, R.D., Zhu, Y.M., Herman, I.P., Steigerwald, M.L., Brus, L.E.: Solid-solution nanoparticles: Use of a nonhydrolytic sol-gel synthesis to prepare HfO_2 and $Hf_xZr_{1-x}O_2$ nanocrystals. Chem. Mater. **16**, 1336–1342 (2004)
136. Tang, J., Redl, F., Zhu, Y., Siegrist, T., Brus, L.E., Steigerwald, M.L.: An organometallic synthesis of TiO_2 nanoparticles. Nano Lett. **5**, 543–548 (2005)
137. Tang, J., Zhang, F., Zoogman, P., Fabbri, J., Chan, S.W., Zhu, Y., Brus, L.E., Steigerwald, M.L.: Martensitic phase transformation of isolated HfO_2, ZrO_2, and $Hf_xZr_{1-x}O_2$ ($0<x<1$) nanocrystals. Adv. Funct. Mater. **15**, 1595–1602 (2005)
138. Tirosh, E., Markovich, G.: Control of defects and magnetic properties in colloidal HfO_2 nanorods. Adv. Mater. **19**, 2608–2612 (2007)
139. Tirosh, E., Shemer, G., Markovich, G.: Optimizing cobalt ferrite nanocrystal synthesis using a magneto-optical probe. Chem. Mater. **18**, 465–470 (2006)
140. Trentler, T.J., Denler, T.E., Bertone, J.F., Agrawal, A., Colvin, V.L.: Synthesis of TiO_2 nanocrystals by nonhydrolytic solution-based reactions. J. Am. Chem. Soc. **121**, 1613–1614 (1999)
141. Varghese, N., Panchakarla, L.S., Hanapi, M., Govindaraj, A., Rao, C.N.R.: Solvothermal synthesis of nanorods of ZnO, N-doped ZnO and CdO. Mater. Res. Bull. **42**, 2117–2124 (2007)
142. Vestal, C.R., Zhang, Z.J.: Effects of interparticle interactions upon the magnetic properties of $CoFe_2O_4$ and $MnFe_2O_4$ nanocrystals. J. Phys. Chem. B **108**, 18,222–18,227 (2004)
143. Wang, D.S., Xie, T., Peng, Q., Zhang, S.Y., Chen, J., Li, Y.D.: Direct thermal decomposition of metal nitrates in octadecylamine to metal oxide nanocrystals. Chem. Eur. J. **14**, 2507–2513 (2007)
144. Wang, H., Uehara, M., Nakamura, H., Miyazaki, M., Maeda, H.: Synthesis of well-dispersed Y_2O_3:Eu nanocrystals and self-assembled nanodisks using a simple non-hydrolytic route. Adv. Mater. **17**, 2506–2509 (2005)
145. Woo, K., Hong, J., Ahn, J.P., Park, J.K., Kim, K.J.: Coordinatively induced length control and photoluminescence of $W_{18}O_{49}$ nanorods. Inorg. Chem. **44**, 7171–7174 (2005)
146. Wood, A., Giersig, M., Mulvaney, P.: Fermi level equilibration in quantum dot-metal nanojunctions. J. Phys. Chem. B **105**, 8810–8815 (2001)
147. Wood, B.D., Mocanu, V., Gates, B.D.: Solution-phase synthesis of crystalline lithium niobate nanostructures. Adv. Mater. **20**, 4552–4556 (2008)
148. Wu, B., Guo, C., Zheng, N., Xie, Z., Stucky, G.D.: Nonaqueous production of nanostructured anatase with high-energy facets. J. Am. Chem. Soc. **130**, 17,563–17,567 (2008)
149. Xia, L.S., Yang, B.F., Fu, Z.P., Yang, Y.L., Yan, H.W., Xu, Y.D., Fu, S.Q., Li, G.P.: High-yield solvothermal synthesis of single-crystalline tin oxide tetragonal prism nanorods. Mater. Lett. **61**, 1214–1217 (2007)
150. Xu, H., Hu, X., Zhang, L.: Generalized low-temperature synthesis of nanocrystalline rare-earth orthoferrites $LnFeO_3$ (Ln = La, Pr, Nd, Sm, Eu, Gd). Cryst. Growth Des. **8**, 2061–2065 (2008)

151. Xu, X., Zhuang, J., Wang, X.: SnO_2 quantum dots and quantum wires: Controllable synthesis, self-assembled 2D architectures, and gas-sensing properties. J. Am. Chem. Soc. **130**, 12,527–12,535 (2008)
152. Yang, H.M., Ouyang, J., Tang, A.D.: Single step synthesis of high-purity CoO nanocrystals. J. Phys. Chem. B **111**, 8006–8013 (2007)
153. Yang, S., Gao, L.: Controlled synthesis and self-assembly of CeO_2 nanocubes. J. Am. Chem. Soc. **128**, 9339–9331 (2006)
154. Yang, T., Shen, C., Li, Z., Zhang, H., Xiao, C., Chen, S., Xu, Z., Shi, D., Li, J., Gao, H.: Highly ordered self-assembly with large area of Fe_3O_4 nanoparticles and the magnetic properties. J. Phys. Chem. B **109**, 23,233–23,236 (2005)
155. Ye, Y., Yuan, F., Li, S.: Synthesis of CoO nanoparticles by esterification reaction under solvothermal conditions. Mater. Lett. **60**, 3175–3178 (2006)
156. Yin, H., Chow, G.: Nonhydrolytic sol-gel synthesis: Microstructural and morphological study on nickel ferrite nanocrystals coated with oleic acid. J. Mater. Res. **23**, 1922–1930 (2008)
157. Yin, M., Gu, Y., Kuskovsky, I.L., Andelman, T., Zhu, Y., Neumark, G.F., O'Brien, S.: Zinc oxide quantum rods. J. Am. Chem. Soc. **126**, 6206–6207 (2004)
158. Yin, M., Wu, C.K., Lou, Y., Burda, C., Koberstein, J.T., Zhu, Y., O'Brien, S.: Copper oxide nanocrystals. J. Am. Chem. Soc. **127**, 9506–9511 (2005)
159. Yu, H., Chen, M., Rice, P.M., Wang, S.X., White, R.L., Sun, S.H.: Dumbbell-like bifunctional Au-Fe_3O_4 nanoparticles. Nano Lett. **5**, 379–382 (2005)
160. Yu, H., Gibbons, P.C., Kelton, K.F., Buhro, W.E.: Heterogeneous seeded growth: A potentially general synthesis of monodisperse metallic nanoparticles. J. Am. Chem. Soc. **123**, 9198–9199 (2001)
161. Yu, T., Joo, J., Park, J., Hyeon, T.: Large-scale nonhydrolytic sol-gel synthesis of uniform-sized ceria nanocrystals with spherical, wire, and tadpole shapes. Angew. Chem. Int. Ed. **44**, 7411–7414 (2005)
162. Yu, T., Joo, J., Park, Y.I., Hyeon, T.: Single unit cell thick samaria nanowires and nanoplates. J. Am. Chem. Soc. **128**, 1786–1787 (2006)
163. Yu, W.W., Falkner, J.C., Yavuz, C.T., Colvin, V.L.: Synthesis of monodisperse iron oxide nanocrystals by thermal decomposition of iron carboxylate salts. Chem. Commun. pp. 2306–2307 (2004)
164. Yuhas, B.D., Zitoun, D.O., Pauzauskie, P.J., He, R., Yang, P.: Transition-metal doped zinc oxide nanowires. Angew. Chem. Int. Ed. **45**, 420–423 (2006)
165. Zeng, H., Li, J., Wang, Z.L., Liu, J.P., Sun, S.H.: Bimagnetic core/shell FePt/Fe_3O_4 nanoparticles. Nano Lett. **4**, 187–190 (2004)
166. Zeng, H., Rice, P.M., Wang, S.X., Sun, S.H.: Shape-controlled synthesis and shape-induced texture of $MnFe_2O_4$ nanoparticles. J. Am. Chem. Soc. **126**, 11,458–11,459 (2004)
167. Zeng, H., Sun, S.H.: Syntheses, properties and potential applications of multicomponent magnetic nanoparticles. Adv. Funct. Mater. **18**, 391–400 (2008)
168. Zhang, H., Edwards, E.W., Wang, D., Möhwald, H.: Directing the self-assembly of nanocrystals beyond colloidal crystallization. Phys. Chem. Chem. Phys. **8**, 3288–3299 (2006)
169. Zhang, H.T., Chen, X.H.: Size-dependent X-ray photoelectron spectroscopy and complex magnetic properties of $CoMn_2O_4$ spinel nanocrystals. Nanotechnology **17**, 1384–1390 (2006)
170. Zhang, H.T., Wu, G., Chen, X.H.: Thermal stability and photoluminescence of $Zr_{1-x}Ce_xO_2$ ($0<x<1$) nanoparticles synthesized in a non-aqueous process. Mater. Chem. Phys. **101**, 415–422 (2007)
171. Zhang, L., Dou, Y.H., Gu, H.C.: Sterically induced shape control of magnetite nanoparticles. J. Cryst. Growth **296**, 221–226 (2006)
172. Zhang, Y.L., Zhong, X.H., Zhu, J., Song, X.: Alcoholysis route to monodisperse CoO nanotetrapods with tunable size. Nanotechnology **18**, 195,605 (2007)

173. Zhang, Z., Liu, S., Chow, S., Han, M.Y.: Modulation of the morphology of ZnO nanostructures via aminolytic reaction: from nanorods to nanosquamas. J. Phys. Chem. B **22**, 6335–6340 (2006)
174. Zhang, Z., Lu, M., Xu, H., Chin, W.S.: Shape-controlled synthesis of zinc oxide: A simple method for the preparation of metal oxide nanocrystals in non-aqueous medium. Chem. Eur. J. **13**, 632–638 (2007)
175. Zhang, Z., Zhong, X., Liu, S., Li, D., Han, M.: Aminolysis route to monodisperse titania nanorods with tunable aspect ratio. Angew. Chem. Int. Ed. **44**, 3466–3470 (2005)
176. Zheng, R.K., Gu, H.W., Xu, B., Fung, K.K., Zhang, X.X., Ringer, S.P.: Self-assembly and self-orientation of truncated octahedral magnetite nanocrystals. Adv. Mater. **18**, 2418–2421 (2006)
177. Zhong, X., Feng, Y., Zhang, Y., Lieberwirth, I., Knoll, W.: Nonhydrolytic alcoholysis route to morphology-controlled ZnO nanocrystals. Small **3**, 1194–1199 (2007)
178. Zhong, X., Xie, R., Sun, L., Lieberwirth, I., Knoll, W.: Synthesis of dumbbell-shaped manganese oxide nanocrystals. J. Phys. Chem. B **110**, 2–4 (2006)
179. Zhong, X.H., Knoll, W.: Morphology-controlled large-scale synthesis of ZnO nanocrystals from bulk ZnO. Chem. Commun. pp. 1158–1160 (2005)
180. Zhou, H.P., Zhang, Y.W., Mai, H.X., Sun, X., Liu, Q., Song, W.G., Yan, C.H.: Spontaneous organization of uniform CeO_2 nanoflowers by 3D oriented attachment in hot surfactant solutions monitored with an in situ electrical conductance technique. Chem. Eur. J. **14**, 3380–3390 (2008)
181. Zhu, P., Wu, W.Y., Zhou, J.P., Zhang, W.: Preparation of size-controlled In_2O_3 nanoparticles. Appl. Organomet. Chem. **21**, 909–912 (2007)
182. Zitoun, D., Pinna, N., Frolet, N., Belin, C.: Single crystal manganese oxide multipods by oriented attachment. J. Am. Chem. Soc. **127**, 15,034–15,035 (2005)

Chapter 4
Solvent-Controlled Synthesis

4.1 Nanoparticles

4.1.1 Introduction

The use of coordinating organic solvents represents an alternative to surfactants, especially in cases where the accessibility of the particle surface as well as the amount of organic impurities are important parameters [42]. In comparison to the synthesis of metal oxides in the presence of surfactants the solvent-controlled approaches are on the one hand considerably simpler (the starting reaction mixture generally just consists of a metal oxide precursor and a common organic solvent), and on the other hand the synthesis temperature is lower, typically in the range of 50 to 200°C. However, without any doubts the main advantage of surfactant-free synthesis methods lies in the improvement of product purity. Surface-adsorbed surfactants not only influence the toxicity of nanoparticles, but also lower the accessibility of the nanoparticle surface in catalytic and sensing applications. These problems are circumvented in nanopowders obtained by surfactant-free routes (cf. also Chapter 2.4).

In the solvent-directed approaches the organic solvent can act as reactant as well as control agent for particle growth and thus makes the use of surfactants superfluous. However, the detailed role of the organic species in these approaches is rather complex. On the one hand they provide the oxygen for the formation of the metal oxide, and on the other hand they act as capping agent, which binds to the particle surface, and thus limits the particle growth, and influences morphology and assembly behavior. The complexity arises from the fact that many organic species may be present in the final synthesis mixture. In addition to the excess of organic solvent and the unreacted organic part of the precursor molecules, also various other organic condensation products are formed as a result of chemical reactions between these organic species. Sometimes the inorganic components, i.e., the metal ion of the precursor molecule or the metal oxide nanoparticles, are able to catalyze the transformation of the organic species into completely unexpected products [99, 94].

Although a rapidly expanding variety of metal oxide nanoparticles has been synthesized by solvent-directed procedures during the last few years [119], only a relatively limited number of reaction systems were used for this purpose [99]. Suitable metal oxide precursors include metal halides, acetates, acetylacetonates, alkoxides, and in cases of more complex oxides also mixtures thereof, and the solvents range from oxygen-containing alcohols, ketones and aldehydes, to oxygen-free solvents like amines, acetonitriles or toluene.

The following sub-chapters discuss the experimental set-ups typically used for nonaqueous and surfactant-free synthesis procedures (4.1.2), followed by the presentation of the most frequently used reaction systems for metal oxide nanoparticle synthesis: i) reaction of metal halides with alcohols (4.1.3), ii) metal alkoxides, acetates and acetylacetonates with alcohols (4.1.4), iii) metal alkoxides with ketones or aldehydes (4.1.5), iv) metal acetylacetonates with amines and nitriles (4.1.6) and v) selected examples of other systems (4.1.7). After the discussion of microwave-supported synthesis approaches to metal oxide nanoparticles (4.1.8), the chapter continues with a section on the synthesis of metal oxide-based organic-inorganic hybrid nanostructures (4.2), and finally ends with a short presentation of the employment of nonaqueous sol-gel techniques in atomic layer deposition (ALD) (4.3).

4.1.2 Experimental Set-up

Depending on the reaction temperature required for the transformation of the molecular precursors into the metal oxide nanocrystals, different heating media can be used (Figure 4.1). If the temperature is well below the boiling point of the organic solvent, then a standard laboratory oil bath on a hot plate combined with stirrer is the easiest option (Figure 4.1a). This set-up is also suitable for experiments under reflux conditions. If higher temperatures, i.e., close or above the boiling points are required, then autoclaves represent the containers of choice. For the production of nanoparticles on a laboratory scale, i.e., 0.1–1 g, typically Parr acid digestion bombs with 45 mL Teflon cups are employed (Figure 4.1b). The Teflon liners provide protection against corrosion. The use of hermetically closed reactors has the advantage of improving the reproducibility by excluding the influence of the surrounding atmosphere. On the other hand, the porosity of the Teflon leads to contamination with nanoparticles, which means that every Teflon cup should only be used for one material. In some selected cases, the role of the autoclave is not only to provide solvothermal conditions or high reaction temperatures, but also to prevent the evaporation of low-boiling organic compounds that need to be present in the reaction vessel to drive nanoparticle formation to completion. For a larger scale, but still laboratory production, it is possible to use 1L autoclaves that offer, in contrast to the smaller ones, mechanical stirring for the homogenization of the reaction mixture. Using this reactor size, it is possible to get metal oxide nanopowders in the range of 20–50 g per batch, as recently shown for ZrO_2 [40], which indicates the promising potential of nonaqueous liquid-phase processes for scalability.

A relatively recent development is the use of microwave irradiation for the synthesis of inorganic nanomaterials. Modern commercial microwave ovens (Figure 4.1c) provide the possibility to stir the reaction mixture, control the temperature and power applied, all of which improve the reproducibility of the synthesis protocol. The direct and efficient heating allows a fast thermal ramping. Reactions are performed in sealed vessels that can stand pressures up to 15 bars. Stirring ensures the homogeneity of the reaction. Rapid cooling can be performed by means of a compressed air jet, so that the reaction can be efficiently thermally quenched. Due to the fact that microwave reactors are able to prevent thermal gradient effects, they bear a great potential for the industrial scale-up of nanoparticle synthesis [47]. The short reaction times ranging from 30 seconds to 3 minutes (cf. Chapter 4.1.8) make it possible to prepare a large number of batches within a short time, thus giving access to large quantities of inorganic nanoparticles with reasonable structural and morphological quality. The syntheses can be performed in 10 or 50 ml glass tubes (Figure 4.1d), sealed with a Teflon cap. This set-up allows a homogeneous penetration of the microwave irradiation into the reaction vessel under precise control of temperature and pressure.

4.1.3 Reaction of Metal Halides with Alcohols

The reaction of metal halides, in general metal chlorides, with alcohols presumably represents the simplest nonaqueous approach to metal oxide nanoparticles, and it can, in the most cases, be performed by simple "beaker" chemistry using an oil bath as heating tool. Metal chlorides like $FeCl_3$ [71], $VOCl_3$ [97, 32], $CoCl_2$ [169], WCl_6 [97, 21, 122, 120], $SnCl_4$ [8], $HfCl_4$ [15], $NbCl_5$ [15], $TaCl_5$ [15] and $TiCl_4$ [152, 96, 92, 1] readily react with various alcohols to the corresponding metal oxides. A particularly well-investigated system is the reaction between $TiCl_4$ and alcohols such as ethanol [152, 92, 153] and benzyl alcohol [97, 96, 1, 154, 173, 2]. In the case of benzyl alcohol, the $TiCl_4$-to-solvent ratio and the reaction temperature influence the crystallite size. Adjustment of these parameters allows the tailoring of the size of the anatase nanocrystals in the range of 4 to 8 nm in small increments [96]. Figure 4.2a-c presents a selection of binary metal oxide nanoparticles obtained from the corresponding metal halides and benzyl alcohol. Whereas the titania nanoparticles exhibit a slightly elongated shape (Figure 4.2a) [96], the hafnium oxide (Figure 4.2b) [15] and tin oxide nanoparticles (Figure 4.2c) [8]

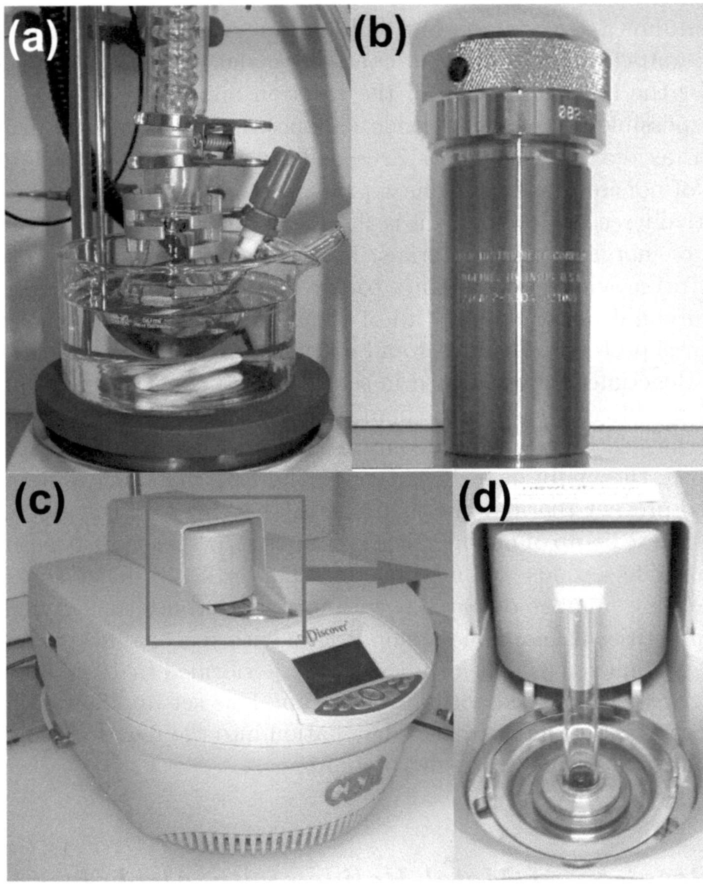

Fig. 4.1. a–d Photographs of a **a** standard laboratory oil bath on a hot plate with stirrer, flask and reflux condenser, **b** Parr autoclave, **c** microwave oven, and **d** microwave reaction tube

are nearly spherical with crystallite sizes of just a few nanometers. An exceptional synthesis system is the reaction between vanadium oxytrichloride with benzyl alcohol, which results in the formation of single-crystalline and semiconducting $VO_{1.52}(OH)_{0.77}$ nanorods with an ellipsoidal morphology, up to 500 nm in length and typically about 100 nm in diameter, and with a hollandite-type crystal structure (Figure 4.2d)[97, 32]. These examples illustrate that the reaction between metal halides and alcohols generally takes place at moderate reaction temperatures, making these systems particularly useful and versatile in cases where organic ligands are required to either functionalize the surface or to modify the crystal morphology.

The low reaction temperature of the $TiCl_4$-benzyl alcohol system (40–150°C) offers the possibility to directly functionalize the surface of the

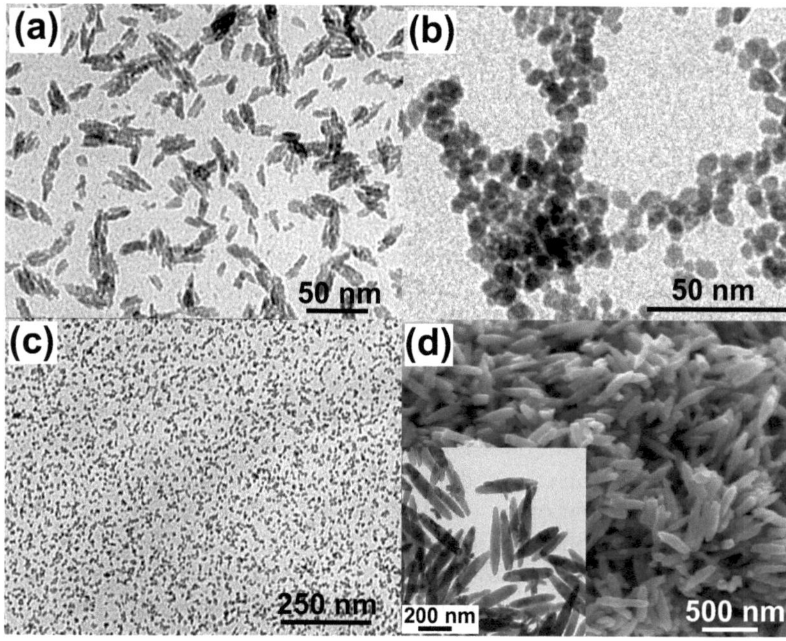

Fig. 4.2. **a–d** TEM images of metal oxide nanoparticles obtained from the corresponding metal chlorides and benzyl alcohol: **a** TiO_2, **b** HfO_2, and **c** SnO_2. **d** SEM image of ellipsoidal $VO_{1.52}(OH)_{0.77}$ nanorods (inset: TEM image) prepared from vanadium oxytrichloride and benzyl alcohol

nanoparticles in an *in situ* process and thus to improve the dispersibility properties [102]. If the preparation of the anatase nanocrystals is performed in the presence of enediol ligands such as dopamine $(HO)_2C_6H_3CH_2CH_2NH_2{\cdot}HCl$ or 4-tert-butylcatechol $(CH_3)_3CC_6H_3(OH)_2$ the final red nanopowders (Figure 4.3a) are well dispersible either in water (Figure 4.3b) or in organic solvents such as THF (Figure 4.3c) depending on the terminal functional group of the ligand [102].

Another example, where the synthesis temperature is low enough to modify the crystal growth and assembly behavior with organic ligands, is nanocrystalline tungsten oxide. The reaction of tungsten chloride with benzyl alcohol yielded tungstite nanoplatelets with a relatively broad size distribution of 30 to 100 nm (Figure 4.3d)[97]. Addition of the bioligand deferoxamine mesylate, a siderophore, changes the particle morphology completely to an anisotropic fibrous-like shape (Figure 4.3e) [122]. At higher magnification it becomes obvious that these fibers consist of individual nanowires that are oriented parallel to each other, and these nanowires are single-crystalline with a uniform diameter of 1.3 nm (Figure 4.3f). If a small amount of 4-tert-butylcatechol is added to the tungsten chloride-benzyl alcohol mixture, anisotropic rod-like architectures with diameters between 35 and 40 nm are observed (Figure 4.3g)[120]. The rods consist of a highly ordered, lamellar organic-inorganic hybrid nanostructure, where the lamellae are stacked in one direction with high precision. The only slightly corrugated sides of the rods prove the good monodispersity of the inorganic building blocks. Not only the addition of organic coordinating species alters the morphological characteristics of the final product, but also the exchange of benzyl alcohol against 4-tert-butylbenzyl alcohol as solvent. The TEM overview image in Figure 4.3h reveals a fibrous-like interconnected network. At higher magnification one can see that these ribbon-like structures consist of parallel columns of uniform diameters of about 4 nm, and each nanocolumn is composed of self-aligned nanoplatelets about 1 nm thick, facing each other along the entire length of the nanostack (Figure 4.3i) [120]. The individual nanostacks are nearly monodisperse in diameter as a consequence of the uniform-sized platelets, which leads to an outstanding alignment on two levels of hierarchy. It is intriguing to see the tremendous difference in crystal growth and in the assembly of the tungsten oxide building blocks, although 4-tert-butylbenzyl alcohol and benzyl alcohol just differ in the presence of a tert-butyl group. Obviously, 4-tert-butylbenzyl alcohol fulfills a number of roles that are directly related to its chemical structure. Similar to benzyl alcohol, the alcohol group provides the oxygen responsible for the formation of tungstite and its stabilizing effect supports the formation of extended flat nanoplatelets. But in contrast to benzyl alcohol, 4-tert-butylbenzyl alcohol with its terminal tertiary butyl group can sterically control both the size of the nanoplatelets and the distance between them. These examples show that the metal halide - alcohol system with its low reaction temperature is particularly useful and versatile in cases where either surface functionalization of the nanoparticles with organic ligands is desired, or where biomimetic reaction principles based on the use of organic templates have to be applied in order to tailor crystal morphology.

4.1.4 Reaction of Metal Alkoxides, Acetates and Acetylacetonates with Alcohols

The reaction of metal halides with alcohols almost always results in metal oxide nanoparticles with halide impurities. To avoid this contamination synthesis routes based on the reaction of metal acetates, acetylacetonates, or alkoxides with alcohols provide a halide-free alternative. Most of the literature is dedicated to the reaction of metal alkoxides with various alcohols, but also a few other examples have been reported. Among them, the so-called polyol route involving the reaction between metal salts and polyols became very popular [37, 38]. In most of these cases, a defined amount of water is added to transform the precursors into the corresponding metal oxides. Using metal acetates as precursors dissolved in polyols, the composition of the final

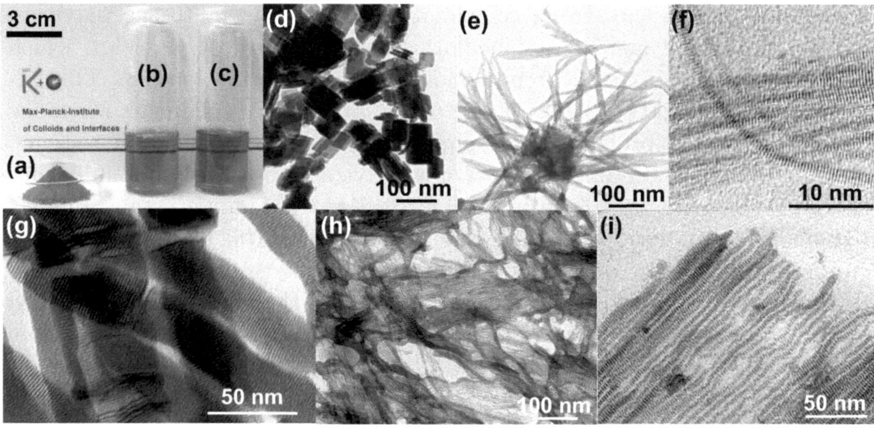

Fig. 4.3. a–i Photograph of **a** dopamine-functionalized titania nanopowder, **b** dispersion of dopamine-functionalized titania in water, **c** dispersion of 4-tert-butylcatechol-functionalized titania in tetrahydrofuran. Image taken from Ref. [97] with permission of the American Chemical Society. TEM overview images of **d** tungsten oxide nanoplatelets obtained from WCl_6 and benzyl alcohol, **e** tungsten oxide nanowire bundles prepared in the presence of deferoxamine mesylate, and **f** HRTEM image of such tungsten oxide nanowires. Images **e** to **f** taken from Ref. [122] with permission of the American Chemical Society. **g** TEM image of stacks of tungsten oxide nanoplatelets synthesized from WCl_6 and benzyl alcohol in the presence of 4-tert-butylcatechol. **h** TEM images of the fibrous-like network of stacks of tungsten oxide nanoplatelets from the reaction of WCl_6 with 4-tert-butylbenzyl alcohol and **i** at higher magnification

material, i.e., whether a metal, metal oxide, or metal hydroxide is formed, depends mainly on the amount of water present in the solution [123]. One example leading to the formation of a metal oxide without any additional water is the reaction between zinc acetate dihydrate and diethyleneglycol at 100–220°C [61]. However, if the synthesis is performed with dehydrated zinc acetate and anhydrous diethyleneglycol, no zinc oxide is obtained. This observation underlines on the one hand the importance and influence of hydrated precursors, and on the other hand shows once more that in some cases it is rather difficult to distinguish between hydrolytic and non-hydrolytic processes. Similarly, $CoFe_2O_4$ nanocrystals of about 4 nm in size were prepared by heating iron and cobalt acetate hydrates in diethyleneglycol at 237°C [5]. CuO nanoparticles in the size range of about 5–10 nm were obtained via alcothermal treatment of copper acetate in ethanol [53]. Solvothermal treatment of iron acetylacetonate in benzyl alcohol resulted in the formation of nanocrystalline magnetite particles with a spherical morphology and diameters of 12–25 nm [117]. The reaction of zirconium n-propoxide in various glycols yielded nanocrystalline zirconia powders at 300°C. The crystallite sizes could be varied in the range of 3 to 5 nm by adjusting the zirconium n-propoxide-to-solvent ratio [86], but the size was also influenced by the nature of the solvent. Whereas 1,2-ethanediol (ethylene glycol) led to the smallest particles due to the strong coordination of the solvent to the particles, 1,4-butanediol (and not 1,6-hexanediol) yielded the largest particles, which was a consequence of the facile intramolecular ether formation [58]. This example nicely highlights the importance of the formation mechanism, when one wants to predict the properties of nanoparticles in a specific reaction system – just considering the binding strength of the glycols is obviously not sufficient. Titanium n-butoxide in 1,4-butanediol formed anatase nanocrystals of about 10-15 nm in diameter after solvothermal reaction [80, 109]. When heating titanium n-butoxide in toluene, only amorphous products were obtained [109]. Ethylene glycol reacts with titanium n-butoxide to titanium glycolate, which can easily be transformed into crystalline anatase nanowires by calcination. Based on this process, SnO_2, In_2O_3 and PbO nanowires with diameters of 50 nm and lengths up to 30 μm were prepared [64]. More information about the glycothermal method can be found somewhere else [56]. In addition to glycols, also benzyl alcohol plays an outstanding role in the nonaqueous preparation of metal oxide nanocrystals from metal alkoxides as precursors. This reaction system is applicable to a large selection of metal oxides such as V_2O_3 [113], Nb_2O_5 [113], Ta_2O_5 [114], HfO_2 [114], SnO_2 [118], In_2O_3 [118], CeO_2 [101, 100], ZrO_2 [40, 172], $NaNbO_3$ [42], $NaTaO_3$ [42], $BaTiO_3$ [104, 103, 12], $LiNbO_3$ [104], $BaZrO_3$ [104], $SrTiO_3$ [103], and (Ba,Sr)TiO_3 [103].

TEM images of nanocrystalline ceria, zirconia, hafnia, indium tin oxide, barium titanate and magnetite are presented in Figure 4.4. In all these cases, the particles exhibit a spherical morphology with uniform shapes, small size distributions and good dispersibility. Ceria, obtained from cerium isopropoxide and benzyl alcohol at 200°C, has an average particle size of 3–4 nm (Figure 4.4a). The reaction of zirconium isopropoxide with benzyl alcohol at 230°C resulted in zirconia nanocrystals with diameters of 4–6 nm (Figure 4.4b).

An important advantage of nonaqueous sol-gel processes, especially in comparison to aqueous systems, is the accessibility of ternary, multi and doped metal oxide nanoparticles [119] (Table 4.1, 4.2 and 4.3). The different reactivity of metal oxide precursors towards a specific solvent complicates the synthesis of oxides containing two or more metals. In organic solvents it is easier to match the reactivity of the metal oxide precursors and of the dopants in comparison to aqueous systems, which is crucial for obtaining single-phase products. In the case of indium tin oxide, the two precursors tin tert-butoxide and indium acetylacetonate were reacted in benzyl alcohol at 200°C, yielding indium tin oxide nanoparticles in the size range of 5–10 nm (Figure 4.4d) [9]. The particularity of this reaction system lies in the simultaneous use of two chemically different precursors. Obviously the reactivity of these precursors in benzyl alcohol matches each other in a way that formation of solid solution takes place, so that the starting In-to-Sn ratio is closely reflected in the final composition of the ITO nanoparticles. This attractive feature enables the controlled preparation of indium tin oxide nanoparticles with varying tin oxide content over the range of 2–30 wt%, and underlines the high potential of nonaqueous sol-gel processes for obtaining doped metal oxide nanoparticles (cf.

also section 4.1.7 and Table 4.3). The maximum electrical conductivity was achieved at a dopant concentration of 15 wt%. Another precursor combination includes cadmium acetate and indium isopropoxide, which results after heating in benzyl alcohol in the formation of $CdIn_2O_4$ nanocrystals with uniform diameters of about 10 nm [18]. The synthesis of $BaTiO_3$ involves the dissolution of metallic barium in benzyl alcohol, addition of titanium isopropoxide and subsequent solvothermal treatment at 200°C for 48 h [104, 103]. This procedure leads to barium titanate nanoparticles with diameters of 5–8 nm (Figure 4.4e). Interestingly, in spite of the small crystal size the structure of these nanoparticles is not completely cubic. Although it is cubic-like on average, it shows locally slight distortions of a tetragonal type [111]. Similarly, $SrTiO_3$ and $(Ba,Sr)TiO_3$ are easily accessible [103].

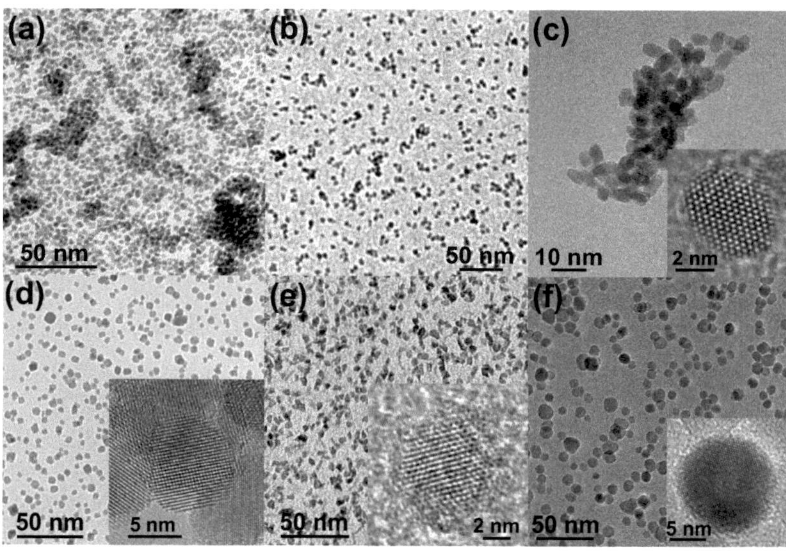

Fig. 4.4. a–f TEM overview images of various metal oxide nanoparticles obtained from halide-free precursors in benzyl alcohol: **a** CeO_2, **b** ZrO_2, **c** HfO_2, **d** SnO_2-doped In_2O_3, **e** $BaTiO_3$, **f** Fe_3O_4

In addition to spherical crystallites, the reaction between metal alkoxides and benzyl alcohol leads in some cases to nanoparticle morphologies with reduced dimensionality. The reaction of niobium ethoxide in benzyl alcohol yielded after solvothermal treatment at 220°C orthorhombic Nb_2O_5 with a platelet-like morphology with sides ranging from 50 to 80 nm (Figure 4.5a) [113]. Bundles of tungsten oxide nanowires are obtained after the reaction of tungsten isopropoxide in benzyl alcohol without the use of any additional structure-directing templates (Figure 4.5b) [121]. The bundles, held together by intercalated benzaldehyde molecules that formed *in situ* through oxidation of benzyl alcohol, can be split up into individual nanowires by the addition of formamide to a dispersion of the nanobundles in ethanol. The small diameter of about 1 nm makes these wires structurally highly flexible (Figure 4.5c), nevertheless, they are well crystalline.

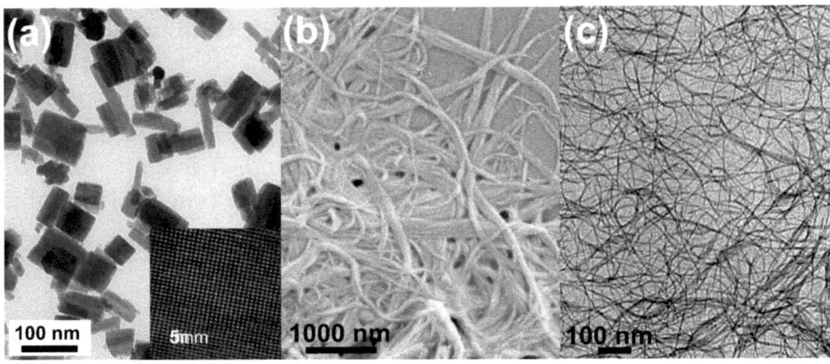

Fig. 4.5. a–c TEM/SEM images of anisotropic nanoparticles obtained from the corresponding metal alkoxides and benzyl alcohol: **a** Nb_2O_5, **b** $W_{18}O_{49}$ nanowire bundles, **c** separated individual $W_{18}O_{49}$ nanowires

4.1.5 Reaction of Metal Alkoxides with Aldehydes and Ketones

Metal oxides containing cations that are sensitive towards reduction to the respective metals by alcohols (in particular copper-, nickel- and lead-containing compounds) are not, or only within a very narrow reaction time window, accessible by the "benzyl alcohol route". In these cases, the use of non-reductive solvents like ketones or aldehydes instead of benzyl alcohol is advantageous. The broad applicability of this approach for the synthesis of these and also other metal oxides has been shown for ZnO [48], TiO_2 [142, 39], In_2O_3 [101] and $BaTiO_3$ [44, 98]. The addition of acetone to benzene solutions of zinc dialkoxides resulted in the formation of transparent gels, which transformed into 4 nm zincite nanocrystals [48]. During this process, acetone is transformed into mesityl oxide via aldolate ligands (β-diketonates) as transition products (cf. Chapter 5.6), which can also act as chelating species, stabilizing the formed nanocrystals. The use of different ketones and aldehydes as oxygen-supplying agents in the nonaqueous, solvothermal synthesis of anatase nanocrystals hence made it possible to tailor the particle size to some extent, as sterically unfavorable alicyclic and aromatic ketones and aldehydes resulted in smaller crystals than aliphatic compounds [39]. Crystalline $BaTiO_3$ was obtained by transformation of the bimetallic alkoxide complex $[(BaTiO)_4(i\text{PrO})_{16}]\cdot 3i\text{PrOH}$ in acetone under stirring at room temperature for 21 days [44]. $BaTiO_3$ was also synthesized in acetophenone starting from metallic barium and titanium isopropoxide [98]. In comparison to $BaTiO_3$ prepared by the "benzyl alcohol route", these nanoparticles are larger, namely in the size range of 10–15 nm. Ketones as solvents were also employed for piezoelectrics such as $PbTiO_3$, $Pb(Zr,Ti)O_3$ and $PbZrO_3$ [41]. These technologically important materials are traditionally synthesized via solid state reactions at temperatures between 600 and 1100°C and nonaqueous sol-gel chemistry offers now a simple soft-chemistry route to amorphous powder precursors that can be transformed into the nanocrystalline oxides at temperatures between 400°C and 600°C [41]. Up to now, also the synthesis of $BaSnO_3$ nanoparticles was only possible in ketones, but not in alcohols [42]. The reaction between iron(III) acetylacetonate, cobalt(III) acetylacetonate and acetophenone yielded cobalt ferrite nanoparticles with diameters from 2–15 nm and the size can be varied in dependence of the reaction temperature [150]. The obtained nanoparticles show superparamagnetic behavior.

4.1.6 Reaction of Metal Acetylacetonates with Amines and Nitriles

In addition to oxygen-containing solvents, also amines and nitriles can be used as solvents for the transformation of metal oxide precursors into binary and ternary metal oxides. It is obvious that in these cases the oxygen for the formation of the oxidic species comes from the precursor molecules, which means that oxygen-containing precursors are a prerequisite for these reaction systems. The most prominent examples are metal acetylacetonates as precursors and benzylamine as solvent. This combination is broadly applicable and gave access to various binary metal oxide nanoparticles such as magnetite/maghemite, γ-Ga_2O_3, zincite ZnO, cubic In_2O_3 [115], as well as the luminescent ternary compound $ZnGa_2O_4$ [17]. The crystal sizes lay in the range of 15–20 nm for the iron, 10–15 nm for the indium, 2.5–3.5 nm for gallium, and around 20 nm for zinc oxide. $ZnGa_2O_4$ was characterized by a very uniform, spherical and rather monodisperse crystal size of 6–8 nm (Figure 4.6a).

Another synthesis system constitutes the solvothermal reaction of metal acetylacetonates with acetonitrile. This approach is characterized by a particularly moderate reaction temperature of 100°C and allows the preparation of nanocrystalline In_2O_3 and ZnO [16]. In this case the oxygen for the metal oxide formation is clearly supplied by the acetylacetonate ligand of the precursor molecules. The In_2O_3 product consists of crystalline nanoparticles with a spherical morphology and a small size distribution in the range of 3–6 nm (Figure 4.6b). According to TEM the ZnO nanoparticles are in the size range of 15–85 nm (Figure 4.6c), sometimes with well-faceted hexagonal morphology (Figure 4.6c, inset).

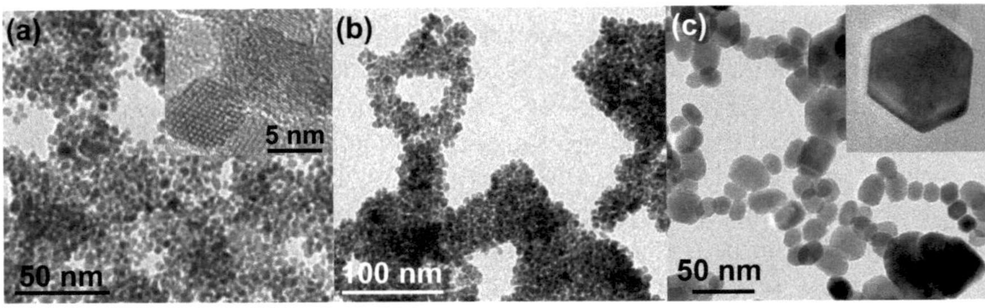

Fig. 4.6. **a**–**c** TEM overview images of **a** $ZnGa_2O_4$ nanocrystals from the metal acetylacetonates and benzylamine (inset: HRTEM image), **b** In_2O_3 from indium acetylacetonate and acetonitrile, and **c** ZnO from zinc acetylacetonate and acetonitrile (inset: TEM image of one hexagon-shaped particle)

4.1.7 Others

Due to the colossal magnetoresistance in doped lanthanum manganites this class of materials has also been targeted using nonaqueous solution routes [149]. However, the lanthanum manganese oxide system is complex and surprising results were reported. The reaction of lanthanum isopropoxide and potassium permanganate in a mixture of the two solvents benzyl alcohol and 2-butanone produced phase-pure lanthanum hydroxide nanofibers (Figure 4.7a and b) [31]. Although $KMnO_4$ is not part of the final inorganic product, it plays a crucial role in determining the particle morphology of $La(OH)_3$. The reason lies in the fact that $KMnO_4$ is able to oxidize the benzyl alcohol to benzoic acid, which influences the anisotropic crystal growth in [001] direction, presumably upon preferential coordination to specific crystal facets of $La(OH)_3$ [31].

A general nonaqueous route for the synthesis of phase-pure transition metal niobate nanocrystals such as $InNbO_4$, $MnNb_2O_6$, and $YNbO_4$ has been reported based on a one-pot solvothermal reaction of niobium chloride and the corresponding transition metal acetylacetonates in benzyl alcohol [163, 164]. Among these niobates, $InNbO_4$ represents a special case, because it is a photocatalytically active ternary metal oxide semiconductor, which works under visible light irradiation. A representative TEM overview image in Figure 4.7c illustrates that the sample consists of nanosized $InNbO_4$ particles, only. Although the particles are slightly agglomerated, due to the lack of any stabilizing surfactants, the grain boundaries are clearly distinguishable. The diameter of the nanoparticles is in the range of 10–30 nm. A HRTEM micrograph of one individual nanoparticle of about 20 nm is shown in Figure 4.7d. The Brunauer-Emmett-Teller (BET) specific surface area was determined to be 54 m^2/g, which is significantly larger than that of its corresponding bulk material (0.67 m^2/g) and might explain its high photocatalytic activity for the degradation of rhodamine B under visible light irradiation (λ

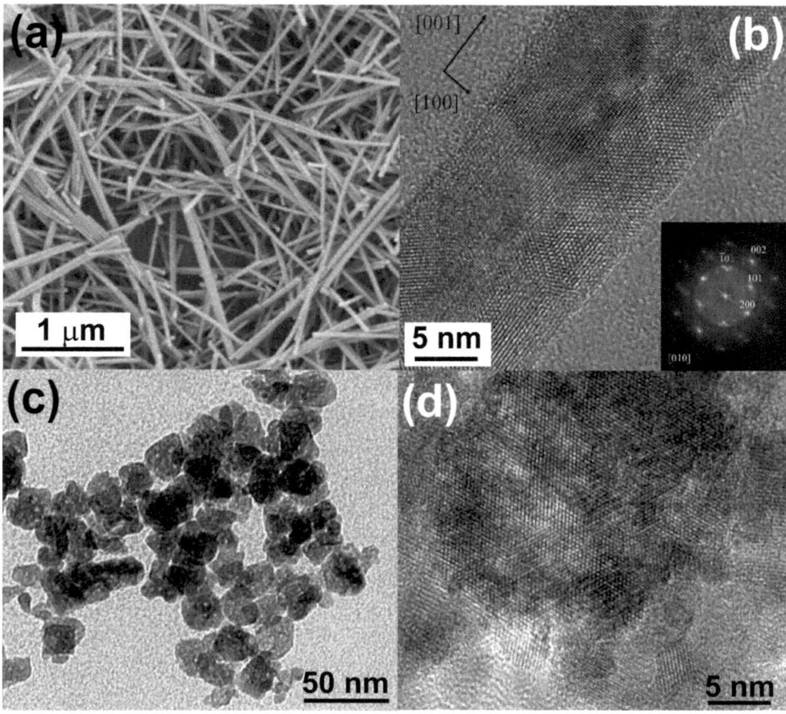

Fig. 4.7. **a** SEM image of $La(OH)_3$ nanofibers, **b** HRTEM of a part of a $La(OH)_3$ (inset: its power spectrum). **c** TEM and **d** HRTEM images of $InNbO_4$ nanoparticles. Images taken from Ref. [163] with permission of Wiley-VCH

$>$ 420 nm) (cf. Chapter 8.4). Nonaqueous reaction conditions seem to be particularly suitable for doping of binary metal oxide nanoparticles as well. The main motivation for these studies is the search for diluted magnetic semiconductors with Curie temperatures above room temperature for applications in spintronics. The most prominent examples include the doping of ZnO with cobalt and manganese [24], manganese-doped ZrO_2 nanocrystals [25], and doping of TiO_2 with iron and cobalt [29]. Table 4.3 gives more examples along these lines.

4.1.8 Microwave Technique

The development of new synthesis routes for inorganic nanomaterials is an integral aspect of materials chemistry. Taking the growing number of nanoparticle-based applications into account also ecological and economical factors are of increasing importance. In this respect, microwave irradiation as a heating tool opens up new perspectives. Microwave-mediated synthesis routes are characterized by short reaction times, often with excellent yields. The high dependency of the crystal size on reaction time, temperature, and concentration, makes it easy to control this parameter through the adjustment of the experimental conditions.

Microwave chemistry is a relatively new field. The first report dates back to 1986 [45]. Although the application of microwave heating is nowadays well known [67], it is almost exclusively applied to the synthesis of organic compounds. However, advantages such as precise control of the temperature, fast thermal ramping, and the direct energy transfer suppressing the wall effect, which is responsible for thermal gradients and inhomogeneous reactions, also offers fascinating opportunities and great benefits to inorganic chemistry, including nanoparticle synthesis. When combined to nonaqueous sol-gel chemistry a large variety of binary and ternary metal oxide nanoparticles can be produced in a time range of 30 s to 20 min in good yields, and thus offers an immense reduction of the reaction times in comparison to traditional heating in an oil bath or in an autoclave. By reacting the corresponding metal alkoxides, acetates or acetylacetonates with benzyl alcohol highly crystalline CoO, ZnO, Fe_3O_4, MnO, Mn_3O_4, and $BaTiO_3$ nanoparticles were obtained (Figure 4.8a) [12]. Representative TEM images show that the magnetite sample in Figure 4.8b (prepared from $Fe(ac)_2$) exhibits a rather uniform spherical particle morphology with a small size distribution in the range of 5 to 10 nm, whereas the $BaTiO_3$ nanoparticles are particularly well dispersed, with sizes in the range of 6-18 nm (Figure 4.8c). All these metal oxides exhibit well-defined lattice fringes in the corresponding HRTEM images (insets in Figure 4.8b and c). Furthermore, it was found that the reaction time as well as the concentration of the precursor have a pronounced effect on the crystal growth, which means that the crystal size can easily be tuned by varying these synthesis parameters [12]. It is worth mentioning that this pronounced dependence of the crystallite size on the heating time was not found in analogous solvothermal experiments in the autoclave.

4.1.9 Tabular Overview of Metal Oxide Nanoparticles

Nowadays, the family of metal oxide nanoparticles that were prepared by nonaqueous and surfactant-free processes has grown immensely and ranges from simple binary metal oxides to more complex ternary and multi metal systems such as perovskites, spinels, and others. Due to space limitations, not all the synthesis routes to the various metal oxide nanoparticles could be discussed in detail. However, Table 4.1 (binary metal oxides), Table 4.2 (ternary and multi metal oxides) and Table 4.3 (doped metal oxides) give a relatively exhaustive overview of the metal oxide nanoparticles synthesized in organic solvents in the absence of water, together with the precursors and the solvents used, the

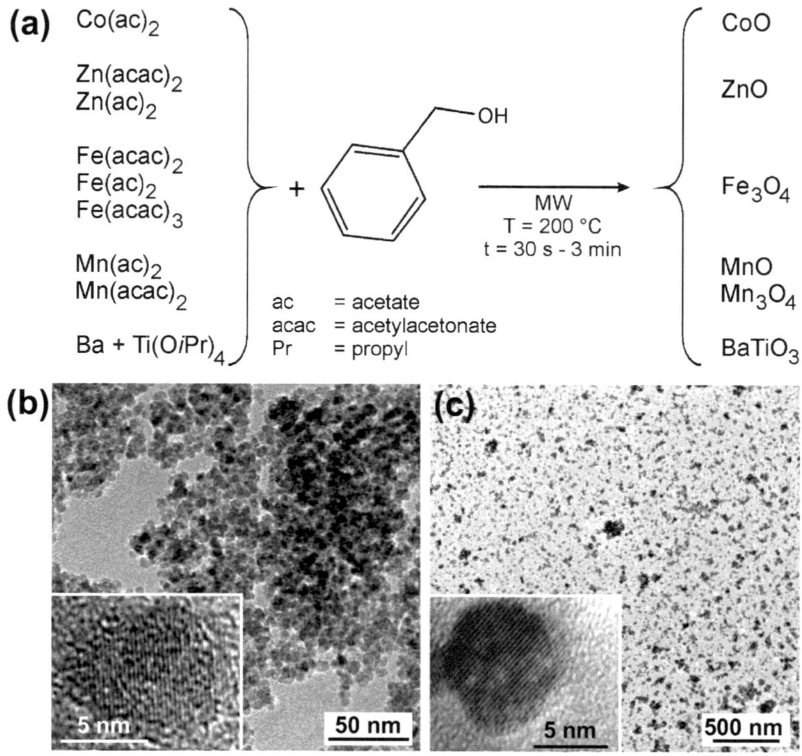

Fig. 4.8. **a** Schematic overview of the different binary and ternary metal oxide nanoparticles produced by the microwave-assisted benzyl alcohol route. **b** TEM and HRTEM images (insets) of **b** Fe_3O_4, and **c** $BaTiO_3$. Images **b** and **c** taken from Ref. [12] with permission of the Royal Society of Chemistry

chemical formation mechanism (if known), and the final crystal morphology (if well-defined). Only liquid phase processes are listed, i.e., thermal decomposition reactions in air or in other gases are not considered. The reaction pathway refers to the condensation step, i.e., the metal-oxygen-metal bond formation, and generally represents the main mechanism in cases, where several parallel pathways were found. Furthermore, one has to keep in mind that some of the precursors are hydrated, which means that the water molecules may be responsible for the hydrolysis and condensation, rendering these approaches hydrolytic. Nevertheless, they were included in the tables under the condition that no additional water was used in the reaction batch.

A closer look at the various precursors and solvents listed in the tables clearly shows that most of the synthesis routes are not generally applicable, but can only be used for a few oxides. One of the exceptions is benzyl alcohol as solvent. It plays an outstanding role, reacting with many metal oxide

Table 4.1. Binary metal oxide nanoparticles synthesized by surfactant-free nonaqueous sol-gel routes; ac = acetate, acac = acetylacetonate

Metal oxide	Precursors	Solvents	Shape	Reaction Mechanism	Ref.
Alumina	Aluminum sec-butoxide	Sec-Butyl alcohol	–	Formation of isobutene and dibutyl ether	[36]
χ-Al_2O_3	Various aluminum alkoxides	Toluene	–	–	[57]
γ-Al_2O_3 or AlOOH	$Al(acac)_3$	Benzylamine, acetophenone, or dimethyl sulfoxide	Spherical (γ-Al_2O_3) and rods (γ-AlOOH)	C-C bond cleavage	[171]
CeO_2	$Ce(acac)_3$	Benzylamine	Rods	–	[100]
CeO_2	$Ce(OiPr)_3$	Benzyl alcohol	Spherical	C-C bond formation	[101, 100]
CeO_2	Ce(III) 2-ethylhexanoate	n-Butylether	Spherical	–	[79]
Co_3O_4	$CoCl_2$	Benzyl alcohol, ethanol, or hexanol	Spherical	–	[169]
CoO	$Co(acac)_2$	Benzyl ether	–	–	[132]
CoO	$Co(ac)_2$	Ethanol	Cube-like	Ester elimination	[162]
CoO	$Co(ac)_2$	Benzyl alcohol	–	–	[12]
Cr_2O_3	$Cr(acac)_3$	1,4-Butanediol	Spherical	–	[125]
Cr_2O_3	$Cr(acac)_3$	Acetone, cyclohexanone or ethanol	Spherical	–	[158]
Cu_2O	$Cu(OEt)_2$	Acetone	–	–	[100]
CuO	$Cu(ac)_2$	Ethanol	Spherical	Ester elimination	[53]
Cu_2O, CuO, ZnO, Al_2O_3, Fe_2O_3, Bi_2O_3, TiO_2, SnO_2, CeO_2, Nb_2O_5, WO_3, $CoFe_2O_4$	Various metal salts, Na_2O	Methanol or ethanol	–	Ion reaction	[20]
Fe_2O_3	$Fe(acac)_3$	1,4-Butanediol, 1-butanol, or toluene	Spherical	–	[124]
Fe_2O_3	$Fe(acac)_3$	Benzylamine	Spherical	C-C bond cleavage	[115]
Fe_3O_4	$Fe(acac)_3$	Benzyl alcohol	Spherical	–	[117]
Fe_2O_3	$Fe(acac)_3$	1,4-Butanediol or toluene	Spherical	–	[19]
Fe_2O_3	$Fe(acac)_3$	2-Acetyl pyridine, p-anisaldehyde, g-butyrolactone, ethylene carbonate, or 1-formyl piperidine	Spherical	–	[158]
Fe_2O_3	$Fe(OC_2H_4OCH_3)_3$	Octadecene	Spherical	–	[3]
Fe_3O_4, Fe_2O_3	$Fe(ac)_2$ or $Fe(ac)_3$	Ethanol, or octanol, or ethanol/acetic acid	Spheres	Ester elimination	[49]
Fe_3O_4	$Fe(acac)_2$, $Fe(ac)_2$ or $Fe(acac)_3$	Benzyl alcohol	–	–	[12]
Fe_3O_4	$FeCl_3{\cdot}H_2O$, sodium acetate	Diethylene glycol	Nanoplates	–	[167]

Table 4.1. (continued)

Metal oxide	Precursors	Solvents	Shape	Reaction Mechanism	Ref.
Ga_2O_3	$Ga(acac)_3$	Benzylamine	Spherical	C-C bond cleavage	[115]
HfO_2	$Hf(OEt)_4$	Benzyl alcohol	Ellipsoidal	Ether elimination	[114]
HfO_2	$HfCl_4$	Benzyl alcohol	Spherical	–	[95]
In_2O_3	$In(OiPr)_3$	Benzyl alcohol	Cube-like	–	[118]
In_2O_3	$In(acac)_3$	Benzylamine	Spherical	C-C bond cleavage	[101, 115]
In_2O_3	$In(OiPr)_3$	Acetophenone	Spherical	–	[101]
In_2O_3	$In(acac)_3$	2-Butanone	Spherical	–	[101]
In_2O_3	$In(acac)_3$	Acetophenone	Spherical	–	[101]
In_2O_3	$In(acac)_3$	Acetonitrile	Spherical	–	[16]
In_2O_3	Indium acetate	Ethanol, ethylene diamine	Nanospheres, nanoflowers and nanorods	Ester elimination	[34]
$La(OH)_3$	$La(OiPr)_3 + KMnO_4$	Benzyl alcohol + 2-butanol	Fibers	Aldol-type reactions	[31]
NiO	$Ni(acac)_2$	1,4-Butanediol	Spherical	–	[125]
Mn_2O_3	$Mn(acac)_3$	1,4-Butanediol	Spherical	–	[125]
Mn_2O_3, Mn_3O_4	$Mn(acac)_3$	Acetone or ethanol	Nearly spherical	–	[158]
MnO and Mn_3O_4	$Mn(acac)_2$ or $KMnO_4$	Benzyl alcohol	Spherical	–	[28]
MnO or Mn_3O_4	$Mn(ac)_2$ or $Mn(acac)_2$	Benzyl alcohol	–	–	[12]
Mn_3O_4	$Mn(ac)_2{\cdot}4H_2O$	KOH + Ethanol	Cube-like	–	[155]
NiO	$Ni(acac)_2$	2-Butanone	Spherical	Keto-enol tautomerization	[11]
ReO_3	$Re_2O_7(C_4H_8O_2)_x$	Toluene	Spherical	–	[13]
SnO_2	$Sn(OtBu)_4$	Benzyl alcohol	Spherical	Ether elimination	[118]
SnO_2	$SnCl_4$	Benzyl alcohol	Spherical	Ether elimination	[8]
SnO_2	Sn(II) 2-ethylhexanoate	n-Butylether	Spherical	–	[79]
Ta_2O_5	$Ta(OEt)_5$	Benzyl alcohol	Spherical	–	[114]
Ta_2O_5	$TaCl_5$	Benzyl alcohol	Spherical	–	[95]
TiO_2	$Ti(OtBu)_4$	p-Xylene, benzene, toluene, or cyclohexane	Spherical	–	[85]
TiO_2	$Ti(OnBu)_4$	Various alcohols	Spherical	–	[84]
TiO_2	$Ti(OiPr)_4$	Various alcohols + formic acid, acetic acid or oxalic acid	–	Ester elimination	[59]
TiO_2	$TiCl_4$	Various alcohols and acetic acid	Spherical or rods	Ester elimination	[152]
TiO_2	$TiCl_4$	Various alcohols	Spherical or rods	Ether elimination	[153]
TiO_2	$TiCl_4$	Benzyl alcohol	Spherical	Alkyl halide elimination	[96, 154, 173] [87, 62]

Table 4.1. (continued)

Metal oxide	Precursors	Solvents	Shape	Reaction Mechanism	Ref.
TiO_2	$Ti(OiPr)_4$	Toluene	Spherical	–	[74]
TiO_2	$Ti(OiPr)_4$	Various ketones	Spherical	Aldol-type reactions	[39]
TiO_2	$Ti(OiPr)_4$	Various aldehydes	Spherical	Aldol-type reactions	[39]
TiO_2	$TiCl_4$	Ethanol	Spherical	–	[92]
TiO_2	$Ti(OnBu)_4$	1,4-Butanediol	Spherical	–	[80]
TiO_2	$Ti(OnBu)_4$	1,4-Butanediol or toluene	Spherical	–	[109]
TiO_2	$Ti(OiPr)_4$	2-Propanol + aniline	Rods	Ester elimination	[170]
TiO_2	Tetrabutyl titanate	n-Butyl alcohol	Spherical	Ether elimination	[161]
TiO_2	$TiCl_4$	Benzyl alcohol or n-butanol	Spherical, slightly elongated	–	[1, 2]
TiO_2 (rutile or anatase)	$TiCl_4$	Acetone and other ketones	Spherical or rods	Aldol condensation	[159, 160]
TiO_2	$Ti(OnBu)_4$	Acetic acid	Spherical	Ester elimination	[63]
TiO_2	$TiBr_4$, Na_2O_2	Hexadecane	–	Metathesis reaction	[110]
V_2O_3	$VO(OiPr)_3$	Benzyl alcohol	–	–	[113]
$VO_{1.52}(OH)_{0.77}$	$VOCl_3$	Benzyl alcohol	Ellipsoids	–	[97, 32]
$WO_3\cdot H_2O$	WCl_6	Benzyl alcohol	Platelets	–	[97]
$W_{18}O_{49}$	WCl_6	Ethanol	Rods	–	[21]
$WO_3\cdot H_2O$	$W(OiPr)_6$	Benzyl alcohol	Wires	–	[121]
$WO_3\cdot H_2O$	WCl_6	4-tert-Butylbenzyl alcohol	Platelets	–	[122]
$W_{18}O_{49}$	WCl_6	Cyclohexanol	Nanowires		[168]
ZnO	$Zn(OCEt_3)_2$	Acetone	Spherical	Aldol-type reactions	[48]
ZnO	$Zn(ac)_2\cdot{}_22H_2O$	Methanol, KOH	Nanorods	–	[107]
ZnO	$Zn(ac)_2\cdot{}_22H_2O$	Ethanol	Spherical	Ester elimination	[33]
ZnO	$Zn(acac)_2$	1,4-Butanediol	Spherical	–	[125]
ZnO	$Zn(acac)_2\cdot xH_2O$	Benzylamine	–	C-C bond cleavage	[115]
ZnO	$Zn(ac)_2$	Benzyl alcohol	Rods or wires	Ester elimination	[24]
ZnO	$Zn(acac)_2\cdot xH_2O$	Acetonitrile	Spherical and hexagonally shaped mesocrystals	–	[16]
ZnO	$Zn(acac)_2\cdot xH_2O$	Benzyl alcohol		–	[12]
ZnO	$Zn(ac)_2\cdot 2H_2O$	1-Pentanol, m-xylene and p-toluene sulfonic acid monohydrate	Spherical	Ester elimination	[27]

Table 4.1. (continued)

Metal oxide	Precursors	Solvents	Shape	Reaction Mechanism	Ref.
ZnO	$Zn(ac)_2$	Various alcohols, glycols, n-alkanes, and aromatic compounds	Various aspect ratios	–	[6]
ZnO	$Zn(acac)_2$	Dibenzylether	–	–	[141]
ZnO	Zinc metal	Aliphatic alcohols	Spherical or nanorods	–	[108]
ZnO	Zinc acetylacetonate hydrate	Benzyl alcohol	–	–	[55]
ZnO	Zinc acetate dihydrate	Diethyleneglycol	Spherical	–	[61]
ZnO	Zinc acetate dihydrate	Diethyleneglycol	Spherical agglomerates	–	[54, 165, 166]
ZrO_2	$Zr(O\mathit{i}Pr)_4{\cdot}HO\mathit{i}Pr$	Glycols or toluene	–	–	[86, 58]
ZrO_2	$Zr(O\mathit{i}Pr)_4{\cdot}HO\mathit{i}Pr$	Benzyl alcohol	Spherical	Ether elimination	[40, 172, 93]

Table 4.2. Ternary and multi metal oxide nanoparticles synthesized by surfactant-free nonaqueous sol-gel routes; ac = acetate, acac = acetylacetonate

Metal oxide	Precursors	Solvents	Shape	Reaction Mechanism	Ref.
$BaSnO_3$	$Ba + Sn(O\mathit{t}Bu)_4$	2-Butanone	Spherical	–	[42]
$BaTiO_3$	$Ba(O\mathit{i}Pr)_2 + Ti(O\mathit{i}Pr)_4$	Acetone	Spherical	Aldol-type reactions	[44]
$BaTiO_3$	$Ba + Ti(O\mathit{i}Pr)_4$	Isopropanol + benzene	–	–	[144]
$BaTiO_3$	$Ba + Ti(O\mathit{i}Pr)_4$	Benzyl alcohol	Spherical	C-C bond formation	[12, 103, 104]
$BaTiO_3$	$Ba + Ti(O\mathit{i}Pr)_4$	Acetophenone	Spherical	Aldol-type reactions	[98]
$BaTiO_3$	$Ba(OH)_2{\cdot}8H_2O + Ti(O\mathit{i}Pr)_4$	2-Methoxyethanol + ethanol	Spherical	–	[82]
$BaTiO_3$	$Ba + Ti(O\mathit{i}Pr)_4$	Isopropanol	–	–	[140]
$BaZrO_3$	$Ba(O\mathit{i}Pr)_2 + Zr(O\mathit{i}Pr)_4{\cdot}HO\mathit{i}Pr$	Acetone	–	–	[44]
$BaZrO_3$	$Ba + Zr(O\mathit{i}Pr)_4{\cdot}HO\mathit{i}Pr$	Benzyl alcohol	Slightly elongated	–	[104]
$(Ba,Sr)TiO_3$	$Ba + Sr + Ti(O\mathit{i}Pr)_4$	Benzyl alcohol	Spherical	–	[103]
$CaNb_2O_6$	$Ca(ac)_2 + Nb(OEt)_5$	1,4-Butanediol	–	–	[83]
$CdIn_2O_4$	$Cd(ac)_2 + In(O\mathit{i}Pr)_3$	Benzyl alcohol	Spherical	–	[18]
$CoFe_2O_4$	$Co(ac)_2 + Fe(ac)_3$	Diethylene glycol	Spherical	–	[10]
$CoFe_2O_4$	$Fe(acac)_3$, $Co(acac)_3$	Acetophenone	Spherical	–	[150]

Table 4.2. (continued)

Metal oxide	Precursors	Solvents	Shape	Reaction Mechanism	Ref.
$CoFe_2O_4$	$Fe(ac)_2$, $Co(ac)_2{\cdot}4H_2O$	Diethyleneglycol	Spherical	–	[5]
$CrNbO_4$	$Cr(acac)_3 + Nb(OnBu)_5$	1,4-Butanediol	–	–	[83]
$FeNbO_3$	$Fe(acac)_3 + Nb(OnBu)_5$	1,4-Butanediol	–	–	[83]
$InNbO_4$	$In(acac)_3 + NbCl_5$	Benzyl alcohol	Spherical	Ether elimination	[163, 164]
Indium Tin Oxide	$In(acac)_3 + Sn(OtBu)_4$	Benzyl alcohol	Spherical	–	[9, 7]
$La_{1-x}A_xMnO_3$ (A = Ca, Sr, Ba)	Various	Benzyl alcohol or acetophenone	–	–	[150]
$LiNbO_3$	$Li(ac) + Nb(OnBu)_5$	1,4-Butanediol	–	–	[83]
$LiNbO_3$	$Li + Nb(OEt)_5$	Benzyl alcohol	–	–	[104]
$LiNbO_3$	LiOH, $Nb(OEt)_5$	1,4-Butanediol	Spherical	Ester elimination	[106]
$LiTaO_3$	LiOH, $Ta(OEt)_5$	1,4-Butanediol	Spherical	Ester elimination	[106]
$MnNb_2O_6$	$Mn(acac)_3 + NbCl_5$	Benzyl alcohol	–	–	[164]
$NaNbO_3$	$Na + Nb(OEt)_5$	Benzyl alcohol	Spherical	–	[42]
$NaTaO_3$	$Na + Ta(OEt)_5$	Benzyl alcohol	–	–	[42]
$PbTiO_3$, $Pb(Zr,Ti)O_3$, $PbZrO_3$	$Pb(acac)_2 + Ti(OiPr)_4$ and/or $Zr(OiPr)_4{\cdot}HOiPr$	2-Butanone	Spherical	–	[41]
RE_3NbO_7 (RE = rare earth)	$RE(ac)_3 + Nb(OEt)_5$	1,4-Butanediol	–	–	[83]
$SrTiO_3$	$Sr + Ti(OiPr)_4$	Benzyl alcohol	Spherical	–	[103]
$SrTiO_3$	$Sr(OiPr)_2 + Ti(OiPr)_4$	Acetone	–	–	[44]
$YNbO_4$	$Y(acac)_3{\cdot}xH_2O + NbCl_5$	Benzyl alcohol	Spherical	–	[164]
ZnM_2O_4 (M=Cr, Fe, Co, Mn)	$Zn(ac)_2 + M(acac)_3$	1,4-Butanediol	Spherical	–	[125]
$ZnGa_2O_4$	$Zn(acac)_2{\cdot}xH_2O$ + $Ga(acac)_3$	Benzylamine	Spherical	–	[17]
$ZnNb_2O_6$	$Zn(ac)_2 + Nb(OnBu)_5$	1,4-Butanediol	–	–	[83]
$Zr_6Nb_2O_{17}$	$Zr(acac)_4 + Nb(OnBu)_5$	1,4-Butanediol	Spherical	–	[83]

Table 4.3. Doped metal oxide nanoparticles synthesized by surfactant-free nonaqueous sol-gel routes

Metal oxide	Precursors	Solvents	Shape	Reaction Mechanism	Ref.
Co-/Fe-doped TiO_2	$Co(acac)_2$, $Fe(acac)_3$, $Ti(O\textit{i}Pr)_4$, $TiCl_4$	Benzyl alcohol	Spherical or well-faceted	–	[29]
Fe-doped TiO_2	$Ti(O\textit{i}Pr)_4 + Fe(ac)_2$	Ethanol and acetic acid	Spherical	Ester elimination	[139]
Co-doped ZnO	$Zn(ac)_2 \cdot 2H_2O$	Ethanol, KOH	Various	–	[156]
Co-doped ZnO	$Zn(ac)_2 + Co(ac)_2$	Benzyl alcohol or benzyl alcohol + anisole	Nanorods and Nanowires	–	[24]
Mn-/Co-doped ZnO	Zn(II) acetylacetonate hydrate, Mn/Co(II) acetylacetonate	Benzyl alcohol	Nanorods	–	[30]
In-doped ZnO	Zinc acetate, indium acetate	Methoxyethanol or ethanol	Spherical	Ester elimination	[22]
Ga-doped ZnO	Zinc acetate, gallium nitrate	Methoxyethanol	Spherical	Ester elimination	[22]
Mn-doped ZrO_2	$Zr(O\textit{i}Pr)_4 \cdot HO\textit{i}Pr$ + $Mn(acac)_3$ or $Mn(ac)_2$	Benzyl alcohol	Spherical	–	[25]

precursors including metal alkoxides, halides, acetylacetonates, or acetates, thus giving access to more than 35 different metal oxide nanoparticles and oxide-based hybrid nanostructures (cf. Chapter 4.2).

The large number of suitable metal oxide precursors and organic solvents offers many combinations for potential reaction systems. Taking into account that the metal oxide precursor as well as the solvent strongly influence the size and the shape of the final nanocrystals, the possibility to chose and to vary the initial reaction mixture represents a powerful tool for the synthetic chemist to tailor the morphological characteristics.

4.2 Organic-Inorganic Hybrid Materials

4.2.1 Introduction

The combination of organic and inorganic components at the nanoscale produces multifunctional materials built up by spatially well defined domains of an organic and an inorganic component [137]. They present a large variety of physical properties arising, not only from the isolated organic and inorganic constituents, but also from the interface formed from the two phases [135]. The definition of the different class of organic-inorganic nanocomposites was drawn by Sanchez and Ribot in the early '90 [136]. In "class I" materials inorganic and organic components interact only weakly through hydrogen bonding, van der Waals contacts or electrostatic forces. In "class II" materials a stronger interaction via ionic or covalent bonds are found. In general, the inorganic species or nanobuilding blocks range from transition metal oxo clusters to polyoxometalates and nanoparticles [137, 134, 73, 72, 138].

The formation of organic-inorganic hybrid nanomaterials in surfactant-free nonaqueous systems was only reported for very few reaction systems. Also in this case, the "benzyl alcohol route" plays an outstanding role, giving access to various oxide based hybrid nanomaterials. The various hybrid materials synthesized in organic solvents are summarized in Table 4.4.

A notable feature of these processes is the fact that a highly ordered hybrid nanostructure is constructed in one-step, governed either by an initially present organic constituent (for example the solvent), or by an organic reaction product formed during the reaction course. The organic species bind strongly and irreversibly to the growing inorganics, fulfilling several functions: i) The selective capping of specific crystal faces leads to a confinement of the inorganic material and the formation of highly anisotropic shapes (e.g. 2D sheets); ii) the organic species serve as stabilizers and prevent agglomeration of the inorganic nanocrystals; iii) the organic species phase-separate from the liquid reaction medium and, e.g. through $\pi-\pi$ interactions, assemble the inorganic nanocrystals into larger superstructures, thereby forming a stable hybrid material.

Table 4.4. Ordered organic-inorganic metal oxide based hybrid nanostructures

Metal oxide	Precursors	Solvents	Structure	Ref.
Y_2O_3	$Y(OiPr)_3$	Benzyl alcohol	Lamellar	[116]
Gd_2O_3	$Gd(OiPr)_3$	Benzyl alcohol	Lamellar	[70]
Nd_2O_3	$Nd(OiPr)_3$	Benzyl alcohol	Lamellar	[70]
Er_2O_3	$Er(OiPr)_3$	Benzyl alcohol	Lamellar	[69]
Sm_2O_3	$Sm(OiPr)_3$	Benzyl alcohol	Lamellar	[70]
Gd_2O_3	$Gd(OiPr)_3$	4-Biphenylmethanol	Lamellar	[68]
Y_2O_3	$Y(OiPr)_3$	4-Biphenylmethanol	Lamellar	[68]
Nd_2O_3	$Nd(OiPr)_3$	4-Biphenylmethanol	Lamellar	[119]
TiO_2	$Ti(OiPr)_4$	Benzylamine	Stacks of nanoplatelets	[43]
$W_{18}O_{49}$	$W(OiPr)_6$	Benzyl alcohol	Nanowire bundles	[121]
$W_{18}O_{49}$	WCl_6 + Deferoxamine	Benzyl alcohol	Nanowire bundles	[122]
$WO_3{\cdot}H_2O$	WCl_6	4-tert-Butylbenzyl alcohol	Stacks of nanoplatelets	[120]
$WO_3{\cdot}H_2O$	WCl_6	Benzyl alcohol + 4-tert-butylcatechol	Stacks of nanoplatelets	[120]

4.2.2 Rare Earth Oxide Based Hybrid Nanoparticles

The reaction of rare earth (RE) alkoxides with benzyl alcohol or a derivative thereof (e.g. 4-biphenylmethanol) in an autoclave between 250-300°C for few days results in the direct formation of ordered lamellar organic-inorganic hybrid materials based on RE_2O_3 oxides [112]. The material consists of very thin crystalline oxide layers of the general formula RE_2O_3, regularly separated from each other by organic layers of intercalated benzoates [116, 70] or biphenolate [68] molecules. They show a typical lamellar structure, composed of i) thin oxide layers which scatter strongly the incident electrons, hence, seen as dark layers on the transmission electron microscopy (TEM) images (Figure 4.9 and 4.10), ii) and an organic part which is practically invisible between those layers. In the case of the materials synthesized in benzyl alcohol a similar structural behavior was found for the different RE. In fact, the interlamellar distance deduced form the TEM pictures was constant at about 1.8 nm and did not depend on the RE. On the other hand, the morphology of the hybrid particles differs in function of the RE. In some cases separated nanoparticles of about 50 nm were found and in other cases they agglomerate like in Figure 4.9b leading to the formation of large agglomerates of primary particles with the lamellar periodicity aligned in the same direction. TEM images of the hybrid materials synthesized in 4-biphenylmethanol instead of benzyl alcohol are shown in Figure 4.10. Similarly to the previ-

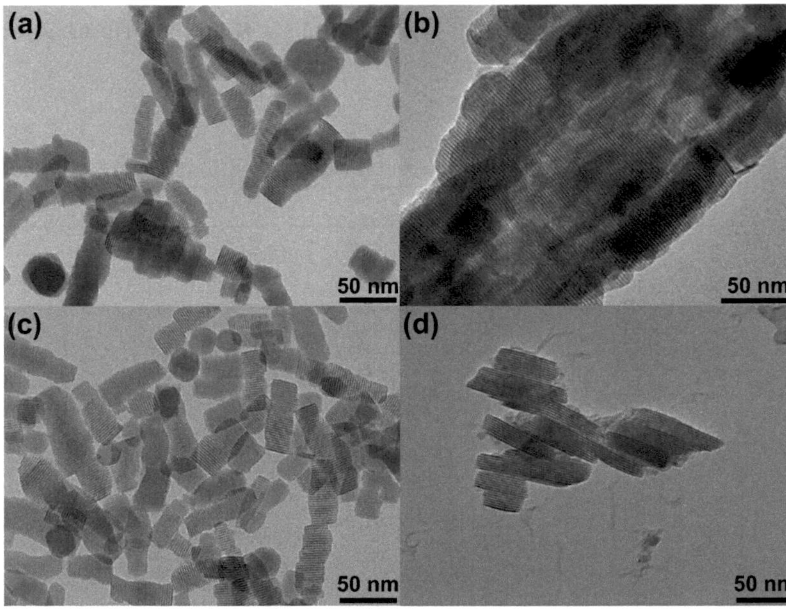

Fig. 4.9. a–d TEM overview images of the **a** yttrium oxide, **b** neodymium oxide, **c** samarium oxide and erbium oxide based benzoate hybrid materials

ous case they show a typical lamellar structure. However, the interlamellar distance increases from 1.8 to 2.6 nm. As demonstrated below, this is due to the larger intercalated molecules. The morphology is also different. On the one hand, gadolinium based materials show less extended oxide sheets compared to the one synthesized in benzyl alcohol (Figure 4.10a). On the other hand, neodymium based hybrid materials show very large oxide sheets that form nanowires several μm long with the lamellar periodicity perpendicular to the long axis (Figure 4.10b). A detailed structural characterization of these nanostructures is not an easy task as single crystals of a reasonable size (~100 μm) could not be obtained. Using several characterization techniques ranging from powder x-ray and electron diffractions, transmission electron microscopy, vibrational spectroscopy, solid state NMR and elemental analysis the structure of the hybrid materials was revealed (cf. Chapter 7.4.2) [116, 70]. It turned out that the oxide layers are composed of only 3 monolayers (~0.6 nm) of crystalline rare earth oxide alternated with a double layer of benzoate or biphenolate species bridged to the oxide nanosheets. A schematic model of the structure is depicted in Figure 4.11. Carboxylate species forming a bridge-like bond with the RE_2O_3 layer were the only organic moieties present between the inorganic layers, and thus they were responsible for the formation of the hybrid structure. In addition, the reaction mixture obtained after the synthesis in benzyl alcohol was carefully investigated in order to

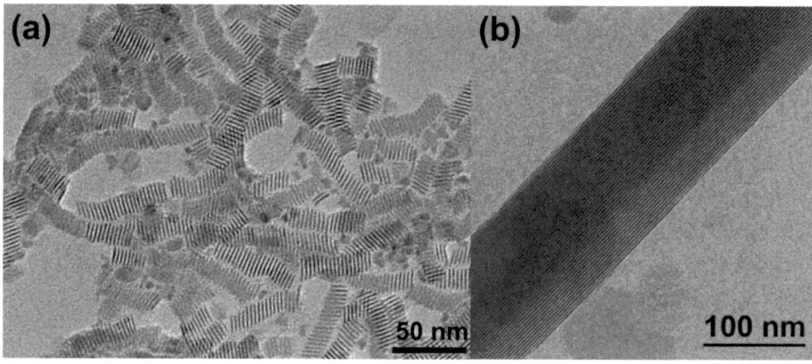

Fig. 4.10. **a**,**b** TEM overview images of the **a** gadolinium oxide, **b** neodymium oxide based biphenolate hybrid materials

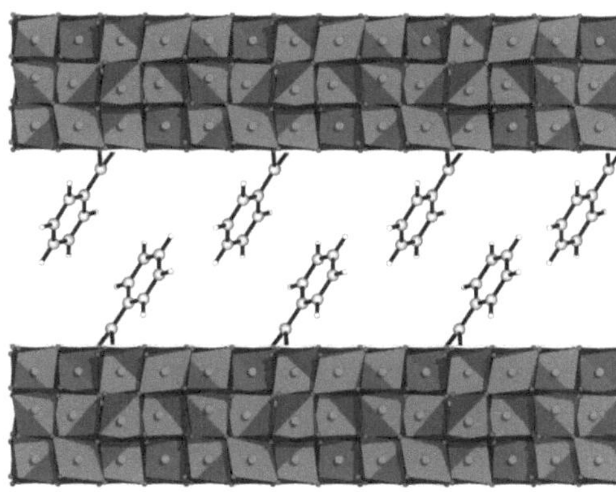

Fig. 4.11 Model of the structure of the benzoate based hybrid material made of three layers of rare earth oxide separated by benzoate species having $\pi-\pi$ interactions

understand how the benzoate species could be formed and why they could control so precisely the growth, and thus the thickness, of the inorganic layer.

Considering the simplicity of the starting reaction mixture (rare earth isopropoxides and benzyl alcohol only) and the complexity of the resulting product, it was clear that the formation must proceed via a number of parallel or consecutive reactions. It was proposed that the formation of the rare earth oxide nanocomposites therefore proceeds via three main coupled reactions [116]: i) A C-C coupling route as recently described for the synthesis of $BaTiO_3$ nanoparticles leading to the elimination of 4-phenyl-2-butanol and formation of a first metal-oxygen-metal bond [103], ii) a disproportionation reaction of benzyl alcohol to toluene and benzaldehyde taking place at the surface of the freshly formed rare earth oxide clusters, and iii) as benzaldehyde was found in much lower proportion than toluene and the nanocomposite contains only benzoate molecules, the rare earth oxide nanostructure catalyses a Cannizaro-like reaction taking place at its surface leading to the formation of the rare earth oxide-benzoate nanocomposite under the elimination of toluene. Both, reaction and structure formation are concerted. The generated surface-bounded carboxylate species stop further structural growth and define the geometry of a lamellar nanostructure of alternating rare earth/benzoate layers, whereas no further benzoate is generated when the surface is saturated. Moreover, the presence of a minimal small-sized rare earth oxide surface is a prerequisite for the surface-catalyzed Cannizaro-like reaction forming benzoate molecules. A detailed explanation of the reaction mechanisms is presented in Chapter 5.5.

4.2.3 Tungsten Oxide Based Hybrid Materials

Another example of hybrid materials is given by the reaction of a tungsten precursor with benzyl alcohol. This system is very rich, in fact the structures and morphologies obtained depend strongly on the experimental conditions. When tungsten chloride is reacted in anhydrous benzyl alcohol pure inorganic platelets of tungstite are obtained [97, 122]. A small quantity of a specific ligand, introduced during the synthesis, dramatically affects the morphology and the supramolecular arrangement of the final material. For example, with deferoxamine mesylate the final material is constituted of long wires of $W_{18}O_{49}$. They have a uniform diameter of about 1.3 nm, an aspect ratio of more than 500 and they are organized in bundles (Figure 4.12a). Similar nanowires were also obtained without additional ligands when tungsten isopropoxide was used instead of tungsten chloride [121]. The bundles were kept together by $\pi - \pi$ interactions between benzaldehyde molecules (oxidation product of benzyl alcohol) adsorbed at their surface [122], forming however much weaker bonds to the tungsten oxide than the carboxylate species did with the rare earth oxides presented in the previous section. Figure 4.12c and d propose a structural model for the tungsten oxide nanowires and their assembly behavior resulting in the hybrid nanostructure. When 4-tert-butylcatechol is used as ligand, long rods consisting of highly ordered lamellar organic-inorganic hybrid structures similar to the ones found for the rare earth hybrid materials are observed [120] (Figure 4.12b).

4.2.4 Hybrid Materials Synthesized in Other Solvents

The reaction of various metal acetylacetonates in benzylamine was previously introduced (cf. Chapter 4.1.6) to be a general approach toward the synthesis of pure metal oxide nanoparticles [115]. The use of different metal complexes permits in certain cases to obtain organic-inorganic materials. The solvothermal treatment of titanium isopropoxide in benzylamine (or m-

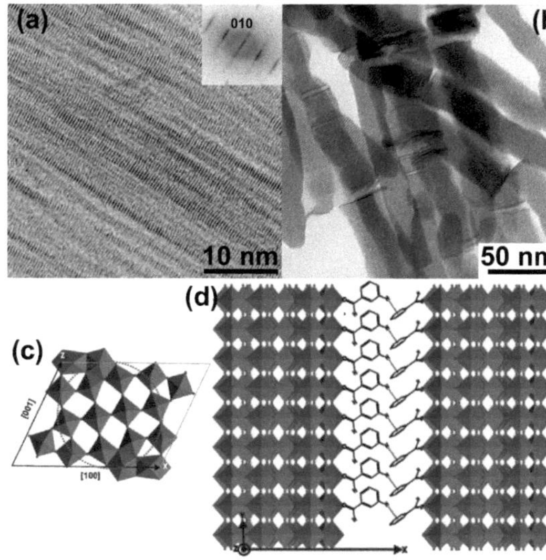

Fig. 4.12 a–d Tungsten oxide hybrid nanostructures synthesized with **a** deferoxamine mesylate (inset: its power spectrum) or **b** 4-tert-butylcatechol as ligand. Schemes representing **c** the cross section of a 1.3 nm nanowire inside one $W_{18}O_{49}$ unit cell oriented along the [010] direction and **d** the proposed model for the hybrid nanostructure

xylylenediamine) resulted in the formation of highly ordered hybrid structures [43]. These structures consist of anatase nanoplatelets that were stacked in a lamellar fashion with, in between, a small organic layer made of benzylamine which keeps the superstructure together by $\pi - \pi$ interactions.

It was proposed that the formation of anatase nanoplatelets, in contrast to the bigger and spherical nanoparticles that were obtained in other solvents [39], is a result of strong binding of the benzylamine (or m-xylylenediamine) solvent to the (001) surfaces of the nanocrystals, which thereby prevents further growth in the [001] direction. The multiple role of the solvent benzylamine is not limited to the control of the crystal growth and the formation of the nanostructure. Indeed, the study of the reaction mechanism showed that benzylamine is also indirectly involved in the transformation of titanium isopropoxide into anatase. A self-condensation of benzylamine in dibenzylamine, promoted by titanium centers, leads to metal-coordinated ammonia groups which are unable to induce bridging to another metal center in contrary to hydroxyl groups. As titanium nitride is not formed, ammonia groups are released during the further course of the condensation. The main condensation path was attributed to the β-elimination of the isopropyl ligands resulting in Ti-OH groups that can initiate the formation of oxygen bridges by reacting with other metal centers (cf. Chapter 5.8).

The highest ordering was obtained when the solvothermal treatment was performed at 200°C for 3 d. A typical diffractogram is presented in Figure 4.13. The dominant low-angle reflection at $2\theta = 7.3°$ proves the presence of a highly ordered lamellar structure, representing the 100 reflection of the lamellae. Its position corresponds to a d spacing of 1.2 nm. A second order peak is visible at 14.1°. The signals at larger angles correspond to the crystal structure of the inorganic material and correspond to the anatase phase. The TEM image of the hybrid structure is presented in Figure 4.14. The image indicates the presence of nanoplatelets that are tightly stacked, forming large aggregates several hundred nanometers in size. The stacking is highly regular, with the nanoplatelets additionally being arranged to form extensive layers in their equatorial direction.

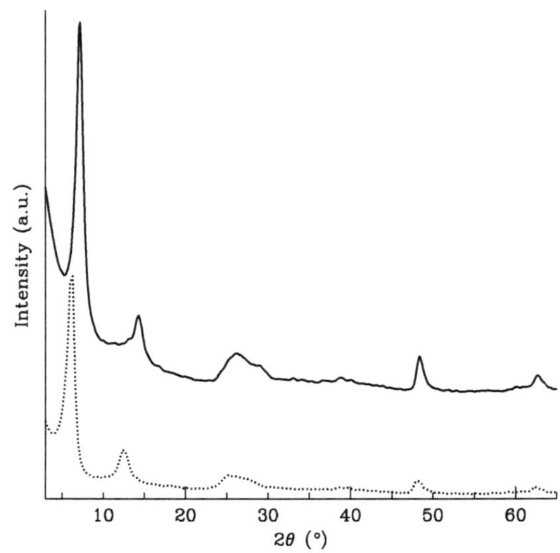

Fig. 4.13 XRD patterns of the nanocomposite obtained from solvothermal treatment of $Ti(OiPr)_4$ in benzylamine (full line) and m-xylylenediamine (dotted line). Image courtesy Prof. Georg Garnweitner

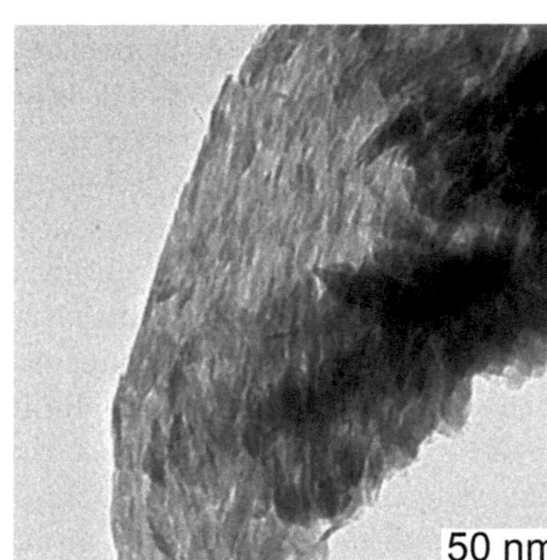

Fig. 4.14 TEM image of the nanocomposite obtained in benzylamine. Image courtesy Prof. Georg Garnweitner

4.3 Nonaqueous Routes Applied to Atomic Layer Deposition

A noble and versatile process for the formation of metal oxide thin films is atomic layer deposition (ALD). ALD is a form of chemical vapor deposition (CVD) in which the reaction between precursor materials is separated into successive surface reactions. In this manner, the precursor materials are kept separate until the adsorbed species react at the surface in a self-limiting process,[143, 105] i.e. without the presence of a gas phase reaction. As a consequence, ALD offers excellent surface conformality. Inherent to the process is the possibility to accurately control the thickness of the deposited film at almost atomic level simply by counting the number of deposition cycles. As an example, Figure 4.15 shows SEM and TEM images of carbon nanotubes homogeneously coated on the inside and the outside by a few nanometers thick titania and hafnia layers [130, 131]. It proves that ALD is probably the most appropriate technique for the coating of complex shapes with high conformality [81].

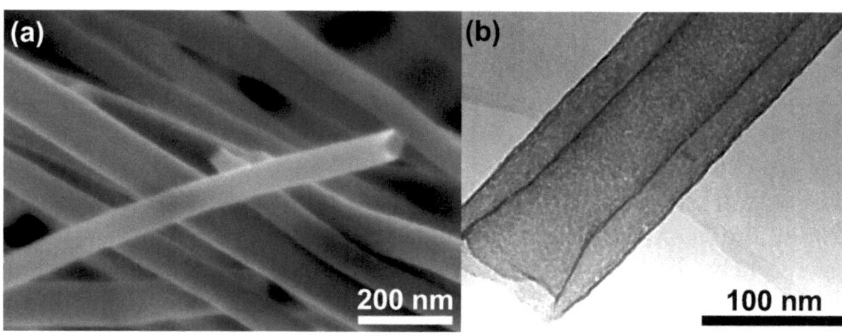

Fig. 4.15. a SEM image of carbon nanotubes coated with titania. **b** TEM image of carbon nanotubes coated with hafnia

Usually, ALD of metal oxides involves the reaction of a metal oxide precursor with an oxygen source. Water is the most commonly used oxygen source and it is generally admitted that water hydrolyzes the surface molecular species which results, at reasonable temperature, in the formation of -OH groups [105, 46, 88]. These react with the metal oxide precursors supplied during the subsequent pulse, leading to the formation of M-O-M bonds. ALD processes using hydrolysable metal complexes and water can therefore be described as successive hydrolysis and condensation steps. Reactions of this kind have been extensively studied in solution and are generally classified as sol-gel chemistry (cf. Chapter 2).

A closer look at the literature reveals that the evolution and strategy developed for the deposition of metal oxide thin films by ALD shows similarities with the development of the sol-gel approach. Both techniques use common precursors such as metal halides, alkoxides and β-diketonates. In ALD, modifications of precursors were also pursued in order to enhance their efficiency by tuning their reactivity/stability [65, 91]. Recently, inspired by the success of nonaqueous sol-gel, similar nonaqueous conditions were applied to ALD [89, 128, 133, 129]. It is also interesting to note that similar chemistry is responsible for the metal oxide formation in ALD and solution, respectively [23].

An exhaustive list of nonaqueous sol-gel approaches applied to ALD is given in Table 4.5. The first example of nonaqueous sol-gel process applied to ALD and also the first mechanistic study found in the literature is attributable to Brei et al. [14]. They alternatively reacted $TiCl_4$ and $Si(OEt)_4$ on a silica surface and monitored the silicate growth by IR and studied the byproducts of the reaction by mass spectrometry. They observed the formation of chloroethane and thereby proved that an alkyl halide elimination mechanism (cf. Chapter 5.2) was involved. Afterward, Ritala et al. [133] showed that the reaction between metal chlorides and alkoxides can be applied for the deposition of various metal oxide and silicate thin films. Up to now, Al_2O_3 [133, 129], Ta_2O_5 [133, 89], Hf [133], Ti [133] and Zr [133] aluminate, $Ti_xHf_yO_z$ [133], $Ti_xZr_yO_z$ [133, 128, 127], Hf [90], Zr [133, 77, 76], and Ti [126, 14] silicate were deposited from their corresponding metal alkoxides and halides. From the data presented in table 4.5 it is clear that the deposition temperatures are relatively high (300-500°C) and definitely higher than the temperatures required to form the metal oxides in solution. In solution the formation of an oxide network typically occurs from room temperature to around 100°C [151, 26, 4] and is extremely fast at higher temperature [66, 148, 146, 145].

More recently, the formation of various metal oxide thin films was achieved by a new approach involving the reaction of a metal alkoxide with carboxylic acids at low temperature (from 50°C) [130]. The proposed reaction mechanism supported by GC-MS and kinetic studies involves an ester elimination (cf. Chapter 5.4) via the following two steps. Firstly, during the carboxylic acid pulse, the replacement of the alkoxy ligand at the surface by a carboxylate group takes place under elimination of alcohol. It is followed, during the metal alkoxide pulse, by a reaction between the carboxylate surface species and the metal alkoxide introduced under elimination of an ester. The as-grown films show an excellent conformality and possess good dielectric properties due to their high purity. Inherent to the chemical approach is the possibility to grow oxides on silicon while minimizing the formation of a low-κ interfacial layer [131].

Metal oxide film formation by ALD under nonaqueous sol-gel conditions did not demonstrate yet significant advantages compared to traditional approaches [23]. However, an interesting aspect is the possibility to deposit metal oxide films without intermediate hydroxyl group formation. Even though the benefit of nonaqueous sol-gel routes in ALD compared to traditional water assisted deposition was carefully studied only for the deposition of alumina [60], it seems that nonaqueous conditions can have a significant impact for the growth of metal oxides on silicon.

In view of what was discussed above, it emerges that metal oxide depositions based on alkyl halide elimination require high deposition temperatures. It is however, not yet clear if this is due to kinetic or thermodynamic reasons [23]. Moreover, in the majority of the cases the contribution of the metal alkoxide self-decomposition (e.g. β-elimination) should be taken into account as it can promote the film growth. Finally, the reaction of metal alkoxides with carboxylic acids, which was only recently reported [130], demonstrated the possibility of a low temperature deposition and self limiting growth even for the deposition of hafnia. Hence, the most promising route appears to be the ester elimination route. It will certainly be the starting point for future studies.

Table 4.5. Metal oxides deposited by ALD using nonaqueous sol-gel approaches

Metal Oxide	Precursor A	Precursor B	Temperature (°C)	Growth per cycle (Å)	Expected condensation step	Ref.
Al_2O_3	$AlCl_3$	$Al(OiPr)_3$	150–375	0.35–1.6	Alkyl halide elimination	[133, 129]
Al_2O_3	$AlCl_3$	$Al(OEt)_3$	400	0.7	Alkyl halide elimination	[133]
Al_2O_3	$AlMe_3$	$Al(OiPr)_3$	300	1.3		[133]
Al_2O_3	$AlMe_3$	HO*i*Pr	250	0.8		[60]
Al_2O_3	$AlCl_3$	HO*t*Bu	300	1.1		[126]
Al_2O_3	$AlCl_3$	HOR	250–500	0.5–0.7		[52, 147]
Al_2O_3	$Al(OR)_3$	HOR	250–500	–		[52]
HfO_2	$Hf(OtBu)_4$	HCOOH	80–200	0.2–0.5	Ester elimination	[130]
HfO_2	$Hf(OtBu)_4$	CH_3COOH	100–350	0.25–1.1	Ester elimination	[130]
Ta_2O_5	$TaCl_5$	$Ta(OEt)_5$	275–450	0.1–0.6	Alkyl halide elimination	[133, 89]
Ta_2O_5	$Ta(OEt)_5$	CH_3COOH	250	0.17	Ester elimination	[130]
TiO_2	$TiCl_4$	HO*t*Bu	200–400	–		[35]
TiO_2	$TiCl_4$	*i*PrO*i*Pr	200–400	0.4–0.6		[35]
TiO_2	$Ti(OiPr)_4$	HCOOH	100–200	0.1–0.5	Ester elimination	[130]
TiO_2	$Ti(OiPr)_4$	CH_3COOH	50–350	0.2–1.4	Ester elimination	[130]
V_2O_4	$VO(OnPr)_5$	CH_3COOH	200	0.4	Ester elimination	[157]
$Al_xHf_yO_z$	$HfCl_4$	$Al(OEt)_3$	400	0.6	Alkyl halide elimination	[133]
$Al_xTi_yO_z$	$AlCl_3$	$Ti(OiPr)_4$	300	2.3	Alkyl halide elimination	[133]
$Al_xZr_yO_z$	$ZrCl_4$	$Al(OEt)_3$	400	0.45	Alkyl halide elimination	[133]
$Ti_xHf_yO_z$	$Hf(OtBu)_4$ $Ti(OiPr)_4$	CH_3COOH	200–300	1.1	Ester elimination	[130]
$Ti_xHf_yO_z$	$HfCl_4$	$Ti(OiPr)_4$	300	0.5	Alkyl halide elimination	[133]
$Ti_xZr_yO_z$	$ZrCl_4$	$Ti(OiPr)_4$	250–300	1.2	Alkyl halide elimination	[133, 128, 127]
Alumina/Silica	$AlCl_3$, $SiBr_4$	HO*t*Bu	300	1.8		[126]
Al silicate/Silica	$AlCl_3$	$(tBuO)_3SiOH$	240	80		[51]
Alumina/Silica	$AlMe_3$	$(tBuO)_3SiOH$	200–300	70–120		[50]
Hf silicate	$HfCl_4$	$Si(nBuO)_4$	300–500	1.4–1.9	No alkyl halide elimination	[75, 78]
Hf silicate/Silica	$HfCl_4$	$(tBuO)_3SiOH$	240			[51]
Hf silicate	$HfCl_4$	$Si(OEt)_4$	500	0.26	Alkyl halide elimination	[90]
Hf silicate	HfI_4	$Si(OEt)_4$	300–500	0.27–0.5	Alkyl halide elimination	[90]
Titania/Silica	$SiCl_4$	$Ti(OtBu)_4$	200–300		Alkyl halide elimination	[126]
Ti silicate	$TiCl_4$	$Si(OEt)_4$	–	–	Alkyl halide elimination	[14]

Table 4.5. (continued)

Metal Oxide	Precursor A	Precursor B	Temperature (°C)	Growth per cycle (Å)	Expected condensation step	Ref.
Zr silicate	$ZrCl_4$	$Si(OEt)_4$	500	0.5	Alkyl halide elimination	[133]
Zr silicate	$ZrCl_4$	$Si(nBuO)_4$	250–500	0.7–1.3	Alkyl halide elimination	[133]
Zr silicate	$ZrCl_4$	$Si(nBuO)_4$	300–500	1.05–1.5	Alkyl halide elimination	[77]
Zr silicate/silica	$ZrCl_4$	$(tBuO)_3SiOH$	240			[51]
Zr silicate	$SiCl_4$	$Zr(OtBu)_4$	125–225	1.1–1.8	Alkyl halide elimination	[76]

The temperatures at which depositions were studied are listed. Note that not necessarily a real ALD process (i.e. self limiting growth) is responsible for film growth over the reported deposition temperatures. The given growth per cycle indicates the minimum and maximum that was obtained over the temperature region studied.

References

1. Abazovic, N.D., Comor, M.I., Dramicanin, M.D., Jovanovic, D.J., Ahrenkiel, S.P., Nedeljkovic, J.M.: Photoluminescence of anatase and rutile TiO_2 particles. J. Phys. Chem. B **110**, 25,366–25,370 (2006)
2. Abazovic, N.D., Ruvarac-Bugarcic, I.A., Comor, M.I., Bibic, N., Ahrenkiel, S.P., Nedeljkovic, J.M.: Photon energy up-conversion in colloidal TiO_2 nanorods. Opt. Mater. **30**, 1139–1144 (2008)
3. Ahniyaz, A., Seisenbaeva, G.A., Häggström, L., Kamali, S., Kessler, V.G., Nordblad, P., Johansson, C., Bergström, L.: Preparation of iron oxide nanocrystals by surfactant-free or oleic acid-assisted thermal decomposition of a Fe(III) alkoxide. J. Magn. Magn. Mater. **320**, 781–787 (2008)
4. Andrianainarivelo, M., Corriu, R., Leclercq, D., Mutin, P.H., Vioux, A.: Mixed oxides SiO_2-ZrO_2 and SiO_2-TiO_2 by a non-hydrolytic sol-gel route. J. Mater. Chem. **6**, 1665–1671 (1996)
5. Artus, M., Ammar, S., Sicard, L., Piquemal, J.Y., Herbst, F., Vaulay, M.J., Fievet, F., Richard, V.: Synthesis and magnetic properties of ferrimagnetic $CoFe_2O_4$ nanoparticles embedded in an antiferromagnetic nio matrix. Chem. Mater. **20**, 4861–4872 (2008)
6. Ayudhya, S.K.N., Tonto, P., Mekasuwandumrong, O., Pavarajarn, V., Praserthdam, P.: Solvothermal synthesis of ZnO with various aspect ratios using organic solvents. Cryst. Growth Des. **6**, 2446–2450 (2006)
7. Ba, J., Feldhoff, A., Fattakhova-Rohlfing, D., Wark, M., Antonietti, M., Niederberger, M.: Crystallization of indium tin oxide nanoparticles: From cooperative behavior to individuality. Small **3**, 310–317 (2007)
8. Ba, J., Polleux, J., Antonietti, M., Niederberger, M.: Nonaqueous synthesis of tin oxide nanocrystals and their assembly into ordered porous mesostructures. Adv. Mater. **17**, 2509–2512 (2005)
9. Ba, J.H., Fattakhova-Rohlfing, D., Feldhoff, A., Brezesinski, T., Djerdj, I., Wark, M., Niederberger, M.: Nonaqueous synthesis of uniform indium tin oxide nanocrystals and their electrical conductivity in dependence of the tin oxide concentration. Chem. Mater. **18**, 2848–2854 (2006)
10. Baldi, G., Bonacchi, D., Franchini, M.C., Gentili, D., Lorenzi, G., Ricci, A., Ravagli, C.: Synthesis and coating of cobalt ferrite nanoparticles: A first step toward the obtainment of new magnetic nanocarriers. Langmuir **23**, 4026–4028 (2007)
11. Beach, E.R., Shqau, K., Brown, S.E., Rozeveld, S.J., Morris, P.A.: Solvothermal synthesis of crystalline nickel oxide nanoparticles. Mater. Chem. Phys. **115**, 371–377 (2009)
12. Bilecka, I., Djerdj, I., Niederberger, M.: One-minute synthesis of crystalline binary and ternary metal oxide nanoparticles. Chem. Commun. pp. 886–888 (2008)
13. Biswas, K., Rao, C.N.R.: Metallic ReO_3 nanoparticles. J. Phys. Chem. B **110**, 842–845 (2006)
14. Brei, V.V., Kaspersky, V.A., Gulyanitskaya, N.U.: Synthesis and study of boron phosphate and titanium silicate compounds on silica surface. Reaction Kinetics and Catalysis Letters **50**, 415–421 (1993)
15. Buha, J., Djerdj, I., Niederberger, M.: Nb, Hf and Ta-chlorides and benzyl alcohol p. unpublished results (2007)
16. Buha, J., Djerdj, I., Niederberger, M.: Nonaqueous synthesis of nanocrystalline indium oxide and zinc oxide in the oxygen-free solvent acetonitrile. Cryst. Growth Des. **7**, 113–116 (2007)
17. Cao, M., Djerdj, I., Antonietti, M., Niederberger, M.: Non-aqueous synthesis of colloidal $ZnGa_2O_4$ nanocrystals and their photoluminescence properties. Chem. Mater. **19**, 5830–5832 (2007)

18. Cao, M., Wang, Y., Chen, T., Antonietti, M., Niederberger, M.: A highly sensitive and fast-responding ethanol sensor based on $CdIn_2O_4$ nanocrystals synthesized by a nonaqueous sol-gel route. Chem. Mater. **20**, 5781–5786 (2008)
19. Chaianansutcharit, S., Mekasuwandumrong, O., Praserthdam, P.: Effect of organic solvents on iron oxide nanoparticles by the solvothermal method. Cryst. Growth Des. **6**, 40–45 (2006)
20. Chen, L., Xu, J., Tanner, D., Phelan, R., der Meulen, M.V., Holmes, J., Morris, M.: One-step synthesis of stoichiometrically defined metal oxide nanoparticles at room temperature. Chem. Eur. J. **15**, 440–448 (2009)
21. Choi, H.G., Jung, Y.H., Kim, D.K.: Solvothermal synthesis of tungsten oxide nanorod/nanowire/nanosheet. J. Am. Ceram. Soc. **88**, 1684–1686 (2005)
22. Cimitan, S., Albonetti, S., Forni, L., Peri, F., Lazzari, D.: Solvothermal synthesis and properties control of doped ZnO nanoparticles. J. Colloid Interface Sci. **329**, 73–80 (2009)
23. Clavel, G., Rauwel, E., Willinger, M.G., Pinna, N.: Nonaqueous sol-gel routes applied to atomic layer deposition of oxides. J. Mater. Chem **19**, 454–462 (2009)
24. Clavel, G., Willinger, M.G., Zitoun, D., Pinna, N.: Solvent dependent shape and magnetic properties of doped ZnO nanostructures. Adv. Funct. Mater. **17**, 3159–3169 (2007)
25. Clavel, G., Willinger, M.G., Zitoun, D., Pinna, N.: Manganese-doped zirconia nanocrystals. Eur. J. Inorg. Chem. pp. 863–868 (2008)
26. Corriu, R.J.P., Leclercq, D., Lefevre, P., Mutin, P.H., Vioux, A.: Preparation of monolithic metal-oxide gels by a non-hydrolytic sol-gel process. J. Mater. Chem. **2**, 673–674 (1992)
27. Demir, M.M., Munoz-Espi, R., Lieberwirth, I., Wegner, G.: Precipitation of monodisperse ZnO nanocrystals via acid-catalyzed esterification of zinc acetate. J. Mater. Chem. **28**, 2940–2947 (2006)
28. Djerdj, I., Arcon, D., Jaglicic, Z., Niederberger, M.: Nonaqueous synthesis of manganese oxide nanoparticles, structural characterization, and magnetic properties. J. Phys. Chem. C **111**, 3614–3623 (2007)
29. Djerdj, I., Arcon, D., Jaglicic, Z., Niederberger, M.: Nonaqueous synthesis of metal oxide nanoparticles: Short review and doped titanium dioxide as case study for the preparation of transition-metal doped oxide nanoparticles. J. Solid State Chem. **181**, 1574–1584 (2008)
30. Djerdj, I., Garnweitner, G., Arcon, D., M., P., Zvonko, J., Niederberger, M.: Diluted magnetic semiconductors: Mn/Co-doped ZnO nanorods as case study. J. Mater. Chem. **18**, 5208–5217 (2008)
31. Djerdj, I., Garnweitner, G., Su, D.S., Niederberger, M.: Morphology-controlled synthesis of anisotropic lanthanum hydroxide nanoparticles. J. Solid State Chem. **180**, 2154–2165 (2007)
32. Djerdj, I., Sheptyakov, D., Gozzo, F., Arcon, D., Nesper, R., Niederberger, M.: Oxygen self-doping in hollandite-type vanadium oxyhydroxide nanorods. J. Am. Chem. Soc. **130**, 11,364–11,375 (2008)
33. Du, H., Yuan, F., Huang, S., Li, J., Zhuy, Y.: A new reaction to ZnO nanoparticles. Chem. Lett. **33**, 770–771 (2004)
34. Du, J., Yang, M., Nam Cha, S., Rhen, D., Kang, M., Kang, D.J.: Indium hydroxide and indium oxide nanospheres, nanoflowers, microcubes, and nanorods: Synthesis and optical properties. Cryst. Growth Des. **8**, 2312–2317 (2008)
35. Evans, P.J., Mutin, P.H., Triani, G., Prince, K.E., Bartlett, J.R.: Characterisation of metal oxide films deposited by non-hydrolytic ALD. Surf. Interface Anal. **38**, 740–743 (2006)
36. Fanelli, A.J., Burlew, J.V.: Preparation of fine alumina powder in alcohol. J. Am. Ceram. Soc. **69**, C174–C175 (1986)
37. Feldmann, C.: Polyol-mediated synthesis of nanoscale functional materials. Adv. Funct. Mater. **13**, 101–107 (2003)

38. Feldmann, C., Jungk, H.O.: Polyol-mediated preparation of nanoscale oxide particles. Angew. Chem. Int. Ed. **40**, 359–362 (2001)
39. Garnweitner, G., Antonietti, M., Niederberger, M.: Nonaqueous synthesis of crystalline anatase nanoparticles in simple ketones and aldehydes as oxygen-supplying agents. Chem. Commun. pp. 397–399 (2005)
40. Garnweitner, G., Goldenberg, L.M., Sakhno, O.V., Antonietti, M., Niederberger, M., Stumpe, J.: Large-scale synthesis of organophilic zirconia nanoparticles and their application in organic-inorganic nanocomposites for efficient volume holography. Small **3**, 1626–1632 (2007)
41. Garnweitner, G., Hentschel, J., Antonietti, M., Niederberger, M.: Nonaqueous synthesis of amorphous powder precursors for nanocrystalline $PbTiO_3$, $Pb(Zr,Ti)O_3$, and $PbZrO_3$. Chem. Mater. **17**, 4594–4599 (2005)
42. Garnweitner, G., Niederberger, M.: Nonaqueous and surfactant-free synthesis routes to metal oxide nanoparticles. J. Am. Ceram. Soc. **89**, 1801–1808 (2006)
43. Garnweitner, G., Tsedev, N., Dierke, H., Niederberger, M.: Benzylamines as versatile agents for the one-pot synthesis and highly ordered stacking of anatase nanoplatelets. Eur. J. Inorg. Chem. pp. 890–895 (2008)
44. Gaskins, B.C., Lannutti, J.J.: Room temperature perovskite production from bimetallic alkoxides by ketone assisted oxo supplementation. J. Mater. Res. **11**, 1953–1959 (1996)
45. Gedye, R., Smith, F., Westaway, K., Ali, H., Baldisera, L., Laberge, L., Rousell, J.: The use of microwave-ovens for rapid organic-synthesis. Tetrahedron Lett. **27**, 279–282 (1986)
46. George, S.M., Ott, A.W., Klaus, J.W.: Surface chemistry for atomic layer growth. J. Phys. Chem. **100**, 13,121–13,131 (1996)
47. Gerbec, J.A., Magana, D., Washington, A., Strouse, G.F.: Microwave-enhanced reaction rates for nanoparticle synthesis. J. Am. Chem. Soc. **127**, 15,791–15,800 (2005)
48. Goel, S.C., Chiang, M.Y., Gibbons, P.C., Buhro, W.E.: New chemistry for the sol-gel process: Acetone as a new condensation reagent. Mater. Res. Soc. Symp. Proc. **271**, 3–13 (1992)
49. Gotic, M., Music, S.: Synthesis of nanocrystalline iron oxide particles in the iron(III) acetate / alcohol / acetic acid system. Eur. J. Inorg. Chem. pp. 966–973 (2008)
50. Hausmann, D., Becker, J., Wang, S.L., Gordon, R.G.: Rapid vapor deposition of highly conformal silica nanolaminates. Science **298**, 402–406 (2002)
51. He, W., Solanki, R., J. F. Conley, J., Ono, Y.: Pulsed deposition of silicate films. J. Appl. Phys. **94**, 3657–3659 (2003)
52. Hiltunen, L., Kattelus, H., Leskela, M., Makela, M., Niinisto, L., Nykanen, E., Soininen, P., Tiittad, M.: Growth and characterization of aluminium oxide thin films deposited from various source materials by atomic layer epitaxy and chemical vapor deposition processes. Mater. Chem. Phys. **28**, 379–388 (1991)
53. Hong, Z.S., Cao, Y., Deng, J.F.: A convenient alcohothermal approach for low temperature synthesis of CuO nanoparticles. Mater. Lett. **52**, 34–38 (2002)
54. Hu, X., Gong, J., Zhang, L., Yu, J.C.: Continuous size tuning of monodisperse ZnO colloidal nanocrystal clusters by a microwave-polyol process and their application for humidity sensing. Adv. Mater. **20**, 4845–4850 (2008)
55. Huang, W., Jiang, P., Wei, C., Zhuang, D., Shi, J.: Low-temperature one-step synthesis of covalently chelated ZnO/dopamine hybrid nanoparticles and their optical properties. J. Mater. Res. **23**, 1946–1952 (2008)
56. Inoue, M.: Glycothermal synthesis of metal oxides. J. Phys.: Condens. Matter **16**, S1291–S1303 (2004)
57. Inoue, M., Kominami, H., Inui, T.: Thermal transformation of χ-alumina formed by thermal decomposition of aluminum alkoxide in organic media. J. Am. Ceram. Soc. **75**, 2597–2598 (1992)

58. Inoue, M., Kominami, H., Inui, T.: Novel synthetic method for the catalytic use of thermally stable zirconia: Thermal decomposition of zirconium alkoxides in organic media. Appl. Catal., A **97**, L25–L30 (1993)
59. Ivanda, M., Music, S., Popovic, S., Gotic, M.: XRD, Raman and FT-IR spectroscopic observations of nanosized TiO_2 synthesized by the sol-gel method based on an esterification reaction. J. Mol. Struct. **481**, 645–649 (1999)
60. Jeon, W.S., Yang, S., Lee, C.s., Kang, S.W.: Atomic layer deposition of Al_2O_3 thin films using trimethylaluminum and isopropyl alcohol. J. Electrochem. Soc. **149**, C306–C310 (2002)
61. Jezequel, D., Guenot, J., Jouini, N., Fievet, F.: Submicrometer zinc oxide particles: Elaboration in polyol medium and morphological characteristics. J. Mater. Res. **10**, 77–83 (1995)
62. Jia, H., Xiao, W.J., Zhang, L., Zheng, Z., Zhang, H., Deng, F.: In situ l-hydroxyproline functionalization and enhanced photocatalytic activity of TiO_2 nanorods. J. Phys. Chem. C **112**, 11,379–11,384 (2008)
63. Jiang, D., Xu, Y., Hou, B., Wu, D., Sun, Y.H.: A simple non-aqueous route to anatase TiO_2. Eur. J. Inorg. Chem. pp. 1236–1240 (2008)
64. Jiang, X., Wang, Y., Herricks, T., Xia, Y.: Ethylene-glycol mediated synthesis of metal oxide nanowires. J. Mater. Chem. **14**, 695–703 (2004)
65. Jones, A., Aspinall, H.C., Chalker, P.R., Potter, R.J., Manning, T.D., Loo, Y.F., O'Kane, R., Gaskell, J.M., Smith, L.M.: MOCVD and ALD of high-κ dielectric oxides using alkoxide precursors. Chem. Vap. Deposition **12**, 83–98 (2006)
66. Joo, J., Yu, T., Kim, Y.W., Park, H.M., Wu, F., Zhang, J.Z., Hyeon, T.: Multigram scale synthesis and characterization of monodisperse tetragonal zirconia nanocrystals. J. Am. Chem. Soc. **125**, 6553–6557 (2003)
67. Kappe, C.O.: Controlled microwave heating in modern organic synthesis. Angew. Chem. Int. Ed. **43**, 6250–6284 (2004)
68. Karmaoui, M., Mafra, L., Sá Ferreira, R.A., Rocha, J., Carlos, L.D., Pinna, N.: Photoluminescent rare-earth based biphenolate lamellar nanostructures. J. Phys. Chem. C **111**, 2539–2544 (2007)
69. Karmaoui, M., Sá Ferreira, R.A., Carlos, L.D., Pinna, N.: Lanthanide-based lamellar nanohybrids: The case of erbium. Mater. Sci. Eng. C **27**, 1368–1371 (2007)
70. Karmaoui, M., Sá Ferreira, R.A., Mane, A.T., Carlos, L.D., Pinna, N.: Lanthanide-based lamellar nanohybrids: Synthesis, structural characterization, and optical properties. Chem. Mater. **18**, 4493–4499 (2006)
71. Khaleel, A.A.: Nanostructured pure γ-Fe_2O_3 via forced precipitation in an organic solvent. Chem. Eur. J. **10**, 925–932 (2004)
72. Kickelbick, G., Holzinger, D., Brick, C., Trimmel, G., Moons, E.: Hybrid inorganic-organic core-shell nanoparticles from surface- functionalized titanium, zirconium, and vanadium oxo clusters. Chem. Mater. **14**, 4382–4389 (2002)
73. Kickelbick, G., Schubert, U.: Inorganic clusters in organic polymers and the use of polyfunctional inorganic compounds as polymerization initiators. Monatsh. Chem. **132**, 13–30 (2001)
74. Kim, C.S., Moon, B.K., Park, J.H., Chung, S.T., Son, S.M.: Synthesis of nanocrystalline TiO_2 in toluene by a solvothermal route. J. Crystal Growth **254**, 405–410 (2003)
75. Kim, M.S., Rogers, S.A., Kim, Y.S., Lee, J.H., Kang, H.K.: Atomic layer deposition analysis of $HfSiO_4$ by mass spectroscopy and XPS. J. Electrochem. Soc. **45**, 1317–1321 (2004)
76. Kim, W.K., Kang, S.W., Rhee, S.W.: Atomic layer deposition of zirconium silicate films using zirconium tetra-tert-butoxide and silicon tetrachloride. J. Vac. Sci. Technol. A **21**, L16–L18 (2003)

77. Kim, W.K., Kang, S.W., Rhee, S.W., Lee, N.I., Lee, J.H., Kang, H.K.: Atomic layer deposition of zirconium silicate films using zirconium tetrachloride and tetra-n-butyl orthosilicate. J. Vac. Sci. Technol. A **20**, 2096–2100 (2002)
78. Kim, W.K., Rhee, S.W., Lee, N.I., Lee, J.H., Kang, H.K.: Atomic layer deposition of hafnium silicate films using hafnium tetrachloride and tetra-n-butyl orthosilicate. J. Vac. Sci. Technol. A **22**, 1285–1289 (2004)
79. Kim, Y.J., Kim, Y.S., Chai, S.Y., Cha, D.H., Choi, Y.S., I.Lee, W.: Syntheses of monodispersed SnO_2 and CeO_2 nanoparticles through the self-capping role of 2-ethylhexanoate ligands. New J. Chem. **31**, 260–264 (2007)
80. Klongdee, J., Petchkroh, W., Phuempoonsathaporn, K., Praserthdam, P., Vangnai, A.S., Pavarajarn, V.: Activity of nanosized titania synthesized from thermal decomposition of titanium (IV) n-butoxide for the photocatalytic degradation of diuron. Sci. Technol. Adv. Mater. **6**, 290–295 (2005)
81. Knez, M., Nielsch, K., Niinisto, L.: Synthesis and surface engineering of complex nanostructures by atomic layer deposition. Adv. Mater. **19**, 3403–3438 (2007)
82. Kolen'ko, Y.V., Kovnir, K.A., Neira, I.S., Taniguchi, T., Ishigaki, T., Watanabe, T., Sakamoto, N., Yoshimura, M.: A novel, controlled, and high-yield solvothermal drying route to nanosized barium titanate powders. J. Phys. Chem. C **111**, 7306–7318 (2007)
83. Kominami, H., Inoue, M., Inui, T.: Formation of niobium double oxides by the glycothermal method. Catal. Today **16**, 309–317 (1993)
84. Kominami, H., Kato, J., Murakami, S., Kera, Y., Inoue, M., Inui, T., Ohtani, B.: Synthesis of titanium IV oxide of ultra-high photocatalytic activity: High-temperature hydrolysis of titanium alkoxides with water liberated homogeneously from solvent alcohols. J. Mol. Catal. A: Chem. **144**, 165–171 (1999)
85. Kominami, H., Kato, J., Takada, Y., Doushi, Y., Ohtani, B., Nishimoto, S., Inoue, M., Inui, T., Kera, Y.: Novel synthesis of microcrystalline titanium(IV) oxide having high thermal stability and ultra-high photocatalytic activity: thermal decomposition of titanium(IV) alkoxide in organic solvents. Catal. Lett. **46**, 235–240 (1997)
86. Kongwudthiti, S., Praserthdam, P., Silveston, P., Inoue, M.: Influence of synthesis conditions on the preparation of zirconia powder by the glycothermal method. Ceram. Int. **29**, 807–814 (2003)
87. Kotsokechagia, T., Cellesi, F., Thomas, A., Niederberger, M., Tirelli, N.: Preparation of ligand-free TIO_2 (anatase) nanoparticles through a nonaqueous process and their surface functionalization. Langmuir **24**, 6988–6997 (2008)
88. Kukli, K., Aarik, J., Aidla, A., Siimon, H., Ritala, M., Leskela, M.: In situ study of atomic layer epitaxy growth of tantalum oxide thin films from $Ta(OC_2H_5)_5$ and H_2O. Appl. Surf. Sci. **112**, 236–242 (1997)
89. Kukli, K., Ritala, M., Leskela, M.: Atomic layer deposition and chemical vapor deposition of tantalum oxide by successive and simultaneous pulsing of tantalum ethoxide and tantalum chloride. Chem. Mater. **12**, 1914–1920 (2000)
90. Kukli, K., Ritala, M., Leskela, M., Sajavaara, T., Keinonen, J., Hegde, R.I., Gilmer, D.C., Tobin, P.J.: Properties of oxide film atomic layer deposited from tetraethoxy silane, hafnium halides, and water. J. Electrochem. Soc. **151**, F98–F104 (2004)
91. Leskela, M., Ritala, M.: Atomic layer deposition (ALD): from precursors to thin film structures. Thin Solid Films **409**, 138–146 (2002)
92. Li, G., Li, L., Boerio-Goates, J., Woodfield, B.F.: High purity anatase TiO_2 nanocrystals: Near room-temperature synthesis, grain growth kinetics, and surface hydration chemistry. J. Am. Chem. Soc. **127**, 8659–8666 (2005)
93. Luo, K., Zhou, S., Wu, L., Gu, G.: Dispersion and functionalization of nonaqueous synthesized zirconia nanocrystals via attachment of silane coupling agents. Langmuir **24**, 11,497–11,505 (2008)

94. Makowski, P., Rothe, R., A., T., Niederberger, M., Goettman, F.: Chlorine borrowing: An efficient method for an easier use of alcohols as alkylation agents. Green Chem. **11**, 34–37 (2009)
95. Niederberger, M., Antonietti, M.: Nanomaterials chemistry: Recent developments and new directions, chap. Nonaqueous sol-gel routes to nanocrystalline metal oxides, pp. 119–138. Wiley-VCH (2007)
96. Niederberger, M., Bartl, M.H., Stucky, G.D.: Benzyl alcohol and titanium tetrachloride: A versatile reaction system for the nonaqueous and low-temperature preparation of crystalline and luminescent titania nanoparticles. Chem. Mater. **14**, 4364–4370 (2002)
97. Niederberger, M., Bartl, M.H., Stucky, G.D.: Benzyl alcohol and transition metal chlorides as a versatile reaction system for the nonaqueous and low-temperature synthesis of crystalline nano-objects with controlled dimensionality. J. Am. Chem. Soc. **124**, 13,642–13,643 (2002)
98. Niederberger, M., Garnweitner, G.: Nonaqueous synthesis of barium titanate nanocrystals in acetophenone as oxygen supplying agent. Mater. Res. Soc. Symp. Proc. **879E**, Z9.8.1–Z9.8.5 (2005)
99. Niederberger, M., Garnweitner, G.: Organic reaction pathways in the nonaqueous synthesis of metal oxide nanoparticles. Chem. Eur. J. **12**, 7282–7302 (2006)
100. Niederberger, M., Garnweitner, G., Ba, J., Polleux, J., Pinna, N.: Nonaqueous synthesis, assembly and formation mechanisms of metal oxide nanocrystals. Int. J. Nanotechnol. **4**, 263–281 (2007)
101. Niederberger, M., Garnweitner, G., Buha, J., Polleux, J., Ba, J., Pinna, N.: Nonaqueous synthesis of metal oxide nanoparticles: overview and indium oxide as case study for the dependence of particle morphology on precursors and solvents. J. Sol-Gel Sci. Technol. **40**, 259–266 (2006)
102. Niederberger, M., Garnweitner, G., Krumeich, F., Nesper, R., Cölfen, H., Antonietti, M.: Tailoring the surface and solubility properties of nanocrystalline titania by a nonaqueous in situ functionalization process. Chem. Mater. **16**, 1202–1208 (2004)
103. Niederberger, M., Garnweitner, G., Pinna, N., Antonietti, M.: Nonaqueous and halide-free route to crystalline $BaTiO_3$, $SrTiO_3$, and $(Ba,Sr)TiO_3$ nanoparticles via a mechanism involving C-C bond formation. J. Am. Chem. Soc. **126**, 9120–9126 (2004)
104. Niederberger, M., Pinna, N., Polleux, J., Antonietti, M.: A general soft chemistry route to perovskites and related materials: Synthesis of $BaTiO_3$, $BaZrO_3$ and $LiNbO_3$ nanoparticles. Angew. Chem. Int. Ed. **43**, 2270–2273 (2004)
105. Niinisto, L., Ritala, M., Leskela, M.: Synthesis of oxide thin films and overlayers by atomic layer epitaxy for advanced applications. Materials Science and Engineering B **41**, 23–29 (1996)
106. Nyman, M., Anderson, T.M., Provencio, P.P.: Comparison of aqueous and nonaqueous soft-chemical syntheses of lithium niobate and lithium tantalate powders. Cryst. Growth Des. **9**, 1036–1040 (2009)
107. Pacholski, C., Kornowski, A., Weller, H.: Self-assembly of ZnO: From nanodots, to nanorods. Angew. Chem. Int. Ed. **41**, 1188–1191 (2002)
108. Panchakarla, L.S., Govindaraj, A., Rao, C.N.R.: Formation of ZnO nanoparticles by the reaction of zinc metal with aliphatic alcohols. J. Cluster Sci. **18**, 660–670 (2007)
109. Payakgul, W., Mekasuwandumrong, O., Pavarajarn, V., Praserthdam, P.: Effects of reaction medium on the synthesis of TiO_2 nanocrystals by thermal decomposition of titanium(IV) n-butoxide. Ceram. Int. **31**, 391–397 (2005)
110. Perera, S., Gillan, E.G.: A facile solvothermal route to photocatalytically active nanocrystalline anatase TiO_2 from peroxide precursors. Solid State Sci. **10**, 864–872 (2008)

111. Petkov, V., Gateshki, M., Niederberger, M., Ren, Y.: Atomic-scale structure of nanocrystalline $Ba_xSr_{1-x}TiO_3$ (x=1, 0.5, 0) by X-ray diffraction and the atomic pair distribution function technique. Chem. Mater. **18**, 814–821 (2006)
112. Pinna, N.: The "benzyl alcohol route": An elegant approach towards organic-norganic hybrid nanomaterials. J. Mater. Chem. **17**, 2769–2774 (2007)
113. Pinna, N., Antonietti, M., Niederberger, M.: A novel nonaqueous route to V_2O_3 and Nb_2O_5 nanocrystals. Colloids Surf., A **250**, 211–213 (2004)
114. Pinna, N., Garnweitner, G., Antonietti, M., Niederberger, M.: Non-aqueous synthesis of high-purity metal oxide nanopowders using an ether elimination process. Adv. Mater. **16**, 2196–2200 (2004)
115. Pinna, N., Garnweitner, G., Antonietti, M., Niederberger, M.: A general non-aqueous route to binary metal oxide nanocrystals involving a C-C bond cleavage. J. Am. Chem. Soc. **127**, 5608–5612 (2005)
116. Pinna, N., Garnweitner, G., Beato, P., Niederberger, M., Antonietti, M.: Synthesis of yttria-based crystalline and lamellar nanostructures and their formation mechanism. Small **1**, 112–121 (2005)
117. Pinna, N., Grancharov, S., Beato, P., Bonville, P., Antonietti, M., Niederberger, M.: Magnetite nanocrystals: Nonaqueous synthesis, characterization, and solubility. Chem. Mater. **17**, 3044–3049 (2005)
118. Pinna, N., Neri, G., Antonietti, M., Niederberger, M.: Nonaqueous synthesis of nanocrystalline semiconducting metal oxides for gas sensing. Angew. Chem. Int. Ed. **43**, 4345–4349 (2004)
119. Pinna, N., Niederberger, M.: Surfactant-free nonaqueous synthesis of metal oxide nanostructures. Angew. Chem. Int. Ed. **47**, 5292–5304 (2008)
120. Polleux, J., Antonietti, M., Niederberger, M.: Ligand and solvent effects in the nonaqueous synthesis of highly ordered anisotropic tungsten oxide nanostructures. J. Mater. Chem. **16**, 3969–3975 (2006)
121. Polleux, J., Gurlo, A., Barsan, N., Weimar, U., Antonietti, M., Niederberger, M.: Template-free synthesis and assembly of single-crystalline tungsten oxide nanowires and their gas-sensing properties. Angew. Chem. Int. Ed. **45**, 261–265 (2006)
122. Polleux, J., Pinna, N., Antonietti, M., Niederberger, M.: Growth and assembly of crystalline tungsten oxide nanostructures assisted by bioligation. J. Am. Chem. Soc. **127**, 15,595–15,601 (2005)
123. Poul, L., Ammar, S., Jouini, N., Fievet, F., Villain, F.: Synthesis of inorganic compounds (metal, oxide and hydroxide) in polyol medium: A versatile route related to the sol-gel process. J. Sol-Gel Sci. Technol. **26**, 261–265 (2003)
124. Praserthdam, P., Mekasuwandumrong, O., Phungphadung, J., Kanyanucharat, A.: New correlation for the effects of the crystallite size and calcination temperature on the single iron oxide nanocrystallites. Cryst. Growth Des. **3**, 215–219 (2003)
125. Praserthdam, P., Silveston, P.L., Mekasuwandumrong, O., Pavarajarn, V., Phungphadung, J., Somrang, P.: A new correlation for the effects of the crystallite size and calcination temperature on the single metal oxides and spinel oxides nanocrystal. Cryst. Growth Des. **4**, 39–43 (2004)
126. Prince, K.E., Evans, P.J., Triani, G., Zhang, Z., Bartlett, J.: Characterisation of alumina-silica films deposited by ALD. Surface and Interface Analysis **38**, 1692–1695 (2006)
127. Rahtu, A., Ritala, M.: Reaction mechanism studies on the atomic layer deposition of $Zr_xTi_yO_z$ using the novel metal halide-metal alkoxide approach. Langmuir **18**, 10,046–10,048 (2002)
128. Rahtu, A., Ritala, M., Leskela, M.: Atomic layer deposition of zirconium titanium oxide from titanium isopropoxide and zirconium chloride. Chem. Mater. **13**, 1528–1532 (2001)

129. Raisanen, P.I., Ritala, M., Leskela, M.: Atomic layer deposition of Al_2O_3 films using $AlCl_3$ and $Al(OiPr)_3$ as precursors. J. Mater. Chem. **12**, 1415–1418 (2002)
130. Rauwel, E., Clavel, G., Willinger, M.G., Rauwel, P., Pinna, N.: Non-aqueous routes to metal oxide thin films by atomic layer deposition. Angew. Chem., Int. Ed. **47**, 3592–3595 (2008)
131. Rauwel, E., Willinger, M.G., Ducroquet, F., Rauwel, P., Matko, I., Kiselev, D., Pinna, N.: Carboxylic acids as oxygen sources for the atomic layer deposition of high-κ metal oxides. J. Phys. Chem. C **112**, 12,754–12,759 (2008)
132. Risbud, A.S., Snedeker, L.P., Elcombe, M.M., Cheetham, A.K., Seshadri, R.: Wurtzite CoO. Chem. Mater. **17**, 834–838 (2005)
133. Ritala, M., Kukli, K., Rahtu, A., Raisanen, P.I., Leskela, M., Sajavaara, T., Keinonen, J.: Atomic layer deposition of oxide thin films with metal alkoxides as oxygen sources. Science **288**, 319–321 (2000)
134. Rozes, L., Steunou, N., Fornasieri, G., Sanchez, C.: Titanium-oxo clusters, versatile nanobuilding blocks for the design of advanced hybrid materials. Monatsh. Chem. **137**, 501–528 (2006)
135. Sanchez, C., Julian, B., Belleville, P., Popall, M.: Applications of hybrid organic-inorganic nanocomposites. J. Mater. Chem. **15**, 3559–3592 (2005)
136. Sanchez, C., Ribot, F.: Design of hybrid organic-inorganic materials synthesized via sol-gel chemistry. New J. Chem. **18**, 1007–1047 (1994)
137. Sanchez, C., Soler-Illia, G., Ribot, F., Lalot, T., Mayer, C.R., Cabuil, V.: Designed hybrid organic-inorganic nanocomposites from functional nanobuilding blocks. Chem. Mater. **13**, 3061–3083 (2001)
138. Schubert, U.: Organically modified transition metal alkoxides: Chemical problems and structural issues on the way to materials syntheses. Acc. Chem. Res. **40**, 730–737 (2007)
139. Sijakovic-Vujicic, N., Gotic, M., Music, S., Ivanda, M., Popovic, S.: Synthesis and microstructural properties of Fe-TiO_2 nanocrystalline particles obtained by a modified sol-gel method. J. Sol-Gel Sci. Technol. **30**, 5–19 (2004)
140. Smith, M.B., Page, K., Siegrist, T., Redmond, P.L., Walter, E.C., Seshadri, R., Brus, L.E., Steigerwald, M.L.: Crystal structure and the paraelectric-to-ferroelectric phase transition of nanoscale $BaTiO_3$. J. Am. Chem. Soc. **130**, 6955–6963 (2008)
141. Snedeker, L.P., Risbud, A.S., Masala, O., Zhang, J.P., Seshadri, R.: Organic phase conversion of bulk (wurtzite) ZnO to nanophase (wurtzite and zinc blende) ZnO. Solid State. Sci. **7**, 1500–1505 (2005)
142. Steunou, N., Ribot, F., Boubekeur, K., Maquet, J., Sanchez, C.: Ketones as an oxolation source for the synthesis of titanium-oxo-organo clusters. New J. Chem. **23**, 1079–1086 (1999)
143. Suntola, T.: Surface chemistry of materials deposition at atomic layer level. Appl. Surf. Sci. **100-101**, 391–398 (1996)
144. Takeuchi, T., Tabuchi, M., Ado, K., Honjo, K., Nakamura, O., Kageyama, H., Suyama, Y., Ohtori, N., Nagasawa, M.: Grain size dependence of dielectric properties of ultrafine $BaTiO_3$ prepared by a sol-crystal method. J. Mater. Sci. **32**, 4053–4060 (1997)
145. Tang, J., Fabbri, J., Robinson, R.D., Zhu, Y.M., Herman, I.P., Steigerwald, M.L., Brus, L.E.: Solid-solution nanoparticles: Use of a nonhydrolytic sol-gel synthesis to prepare HfO_2 and $Hf_xZr_{1-x}O_2$ nanocrystals. Chem. Mater. **16**, 1336–1342 (2004)
146. Tian, B., Yang, H., Liu, X., Xie, S., Yu, C., Fan, J., Tu, B., Zhao, D.: Fast preparation of highly ordered nonsiliceous mesoporous materials via mixed inorganic precursors. Chem. Commun. pp. 1824–1825 (2002)
147. Tiitta, M., Nykanen, E., Soininen, P., Niinisto, L., Leskela, M., Lappalainen, R.: Preparation and characterization of phosphorus-doped aluminum oxide thin films. Materials Research Bulletin **33**, 1315–1323 (1998)

148. Trentler, T.J., Denler, T.E., Bertone, J.F., Agrawal, A., Colvin, V.L.: Synthesis of TiO_2 nanocrystals by nonhydrolytic solution-based reactions. J. Am. Chem. Soc. **121**, 1613–1614 (1999)
149. Vazquez-Vazquez, C., Lopez-Quintela, M.A.: Solvothermal synthesis and characterisation of $La_{1-x}A_xMnO_3$ nanoparticles. J. Solid State Chem. **179**, 3229–3237 (2006)
150. Vazquez-Vazquez, C., Lovelle, M., Mateo, C., Lopez-Quintela, M.A., Bujan-Nunez, M.C., Serantes, D., Baldomir, D., Rivas, J.: Magnetocaloric effect and size-dependent study of the magnetic properties of cobalt ferrite nanoparticles prepared by solvothermal synthesis. Phys. Stat. Sol. A **205**, 1358–1362 (2008)
151. Vioux, A.: Nonhydrolytic sol-gel routes to oxides. Chem. Mater. **9**, 2292–2299 (1997)
152. Wang, C., Deng, Z., Li, Y.: The synthesis of nanocrystalline anatase and rutile titania in mixed organic media. Inorg. Chem. **40**, 5210–5214 (2001)
153. Wang, C., Deng, Z.X., Zhang, G.H., Fan, S.S., Li, Y.D.: Synthesis of nanocrystalline TiO_2 in alcohols. Powder Technol. **125**, 39–44 (2002)
154. Wang, J., Polleux, J., Lim, J., Dunn, B.: Pseudocapacitive contributions to electrochemical energy storage in TiO_2 (anatase) nanoparticles. J. Phys. Chem. C **111**, 14,925–14,931 (2007)
155. Wang, N., Guo, L., He, L., Cao, X., Chen, C., Wang, R., Yang, S.: Facile synthesis of monodisperse Mn_3O_4 tetragonal nanoparticles and their large-scale assembly into highly regular walls by a simple solution route. Small **3**, 606–610 (2007)
156. Wang, X.F., Xu, J.B., Zhang, B., Yu, H.G., Wang, J., Zhang, X., Yu, J.G., Li, Q.: Signature of intrinsic high-temperature ferromagnetism in cobalt-doped zinc oxide nanocrystals. Adv. Mater. **18**, 2476–2480 (2006)
157. Willinger, M., Neri, G., Rauwel, E., Bonavita, A., Micali, G., Pinna, N.: Vanadium oxide sensing layer grown on carbon nanotubes by a new atomic layer deposition process. Nano Lett. **8**, 4201–4204 (2008)
158. Willis, A.L., Chen, Z.Y., He, J.Q., Zhu, Y.M., Turro, N.J., O'Brien, S.: Metal acetylacetonates as general precursors for the synthesis of early transition metal oxide nanomaterials. J. Nanomater. **2007**, 14,858 (2007)
159. Wu, Y., Liu, H.M., Xu, B.Q.: Solvothermal synthesis of TiO_2: Anatase nanocrystals and rutile nanofibers from $TiCl_4$ in acetone. Appl. Organomet. Chem. **21**, 146–149 (2007)
160. Wu, Y., Liu, H.M., Xu, B.Q., Zhang, Z.L., Su, D.S.: Single-phase titania nanocrystallites and nanofibers from titanium tetrachloride in acetone and other ketones. Inorg. Chem. **46**, 5093–5099 (2007)
161. Yang, X., Konishi, H., Xu, H., Wu, M.: Comparative sol-hydro(solvo)thermal synthesis of TiO_2 nanocrystals. Eur. J. Inorg. Chem. pp. 2229–2235 (2006)
162. Ye, Y., Yuan, F., Li, S.: Synthesis of CoO nanoparticles by esterification reaction under solvothermal conditions. Mater. Lett. **60**, 3175–3178 (2006)
163. Zhang, L., Djerdj, I., Cao, M., Antonietti, M., Niederberger, M.: Nonaqueous sol-gel synthesis of nanocrystalline $InNbO_4$ visible light photocatalyst. Adv. Mater. **19**, 2083–2086 (2007)
164. Zhang, L., Garnweitner, G., Djerdj, I., Antonietti, M., Niederberger, M.: Generalized nonaqueous sol-gel synthesis of different transition metal niobate nanocrystals and analysis of the growth mechanism. Chem. Asian J. **3**, 746–752 (2008)
165. Zhang, Q., Chou, T., Russo, B., Jenekhe, S., Cao, G.: Aggregation of ZnO nanocrystallites for high conversion efficiency in dye-sensitized solar cells. Angew. Chem. Int. Ed. **47**, 2402–2406 (2008)
166. Zhang, Q., Chou, T.P., Russo, B., Jenekhe, S.A., Cao, G.: Polydisperse aggregates of ZnO nanocrystallites: A method for energy-conversion-efficiency enhancement in dye-sensitized solar cells. Adv. Funct. Mater. **18**, 1654–1660 (2008)

167. Zhang, W.D., Xiao, H.M., Zhu, L.P., Fu, S.Y.: Template-free solvothermal synthesis and magnetic properties of novel single-crystalline magnetite nanoplates. J. Alloys Compd. **In Print** (2009)
168. Zhao, Y.M., Hu, W.B., Xia, Y.D., Smith, E.F., Zhu, Y.Q., Dunnill, C.W., Gregory, D.H.: Preparation and characterization of tungsten oxynitride nanowires. J. Mater. Chem. **17**, 4436–4440 (2007)
169. Zhao, Z.W., Guo, Z.P., Liu, H.K.: Nonaqueous synthesis of crystalline Co_3O_4 powders using alcohol and cobalt chloride as a versatile reaction system for controllable morphology. J. Power Sources **147**, 264–268 (2005)
170. Zhong, Z., Ang, T.P., Luo, J., Gan, H.C., Gedanken, A.: Synthesis of one-dimensional and porous TiO_2 nanostructures by controlled hydrolysis of titanium alkoxide via coupling with an esterification reaction. Chem. Mater. **17**, 6814–6818 (2005)
171. Zhou, S., Antonietti, M., Niederberger, M.: Low-temperature synthesis of γ-alumina nanocrystals from aluminum acetylacetonate in nonaqueus media. Small **3**, 763–767 (2007)
172. Zhou, S., Garnweitner, G., Niederberger, M., Antonietti, M.: Dispersion behavior of zirconia nanocrystals and their surface functionalization with vinyl group-containing ligands. Langmuir **23**, 9178–9187 (2007)
173. Zhu, J., Yang, J., Bian, Z.F., Ren, J., Liu, Y.M., Cao, Y., Li, H.X., He, H.Y., Fan, K.N.: Nanocrystalline anatase TiO_2 photocatalysts prepared via a facile low temperature nonhydrolytic sol-gel reaction of $TiCl_4$ and benzyl alcohol. Appl. Catal., B **76**, 82–91 (2007)

Chapter 5
Reaction Mechanism

5.1 Introduction

Chemical reactions based on organic chemistry principles play an outstanding, but not yet fully understood role in the nonaqueous synthesis of metal oxide nanoparticles. Nanoparticle formation in organic solvents is always accompanied by organic side reactions that are responsible for the chemical transformation of the molecular precursor to the oxidic compound. In all of them, interplay between the precursor species and the solvent leads to the formal supply of oxygen, but in fact the metal center of the precursor complex acts as strong catalyst for these reactions. Some of these mechanisms are based on common organic condensation reactions that are capable of generating water *in situ*, which then instantly reacts with the metal oxide precursors. In many cases, however, water as such is not formed, but the organic condensation and the formation of metal-oxygen-metal (M-O-M) bridged bonds are concurrent processes [94]. Information on the chemical formation mechanisms can be gained through the identification of the organic by-products, which are either present in the final reaction mixture, or escape during the synthesis as gaseous compounds. By retro-synthetical approaches it is possible to correlate the processes leading to the organic species to the growth of the oxide particles. The correlation of both, the organic reactions to the crystallization of the inorganic nanoparticles, as well as the organic species to nanoparticle morphology represent fundamental and challenging questions. In contrast to aqueous systems with nearly indefinable composition, the characterization of the organic compounds in organic media can easily be performed with standard techniques like Nuclear Magnetic Resonance (NMR) spectroscopy or Gas Chromatography - Mass Spectrometry (GC-MS). In this context, surfactant-free nonaqueous routes are particularly suitable for the study of the reaction pathways due to the small number of starting compounds.

The sol-gel synthesis of metal oxide nanoparticles is based on the formation of M-O-M bonds, representing the first step towards the build-up of the solid metal oxide structure. Obviously, in nonaqueous processes the formal oxygen that is required for the formation of these oxygen bridges is not provided by any added water. Therefore, it must either stem from the precursor used or be provided in some form from the organic medium. In fact, often these cases cannot be clearly distinguished, and the mechanisms for this oxygen transfer are manifold. These mechanisms nonetheless can have a strong influence on the intrinsic properties of the forming solid material, on their composition and the crystal structure, and therefore, their detailed elucidation is vital to bring us closer to understand and to control nanoparticle formation on a molecular level. Central to all of these mechanisms is the formation of the aforementioned M-O-M bonds from the precursors, which through analogous further condensation reactions grow into larger 3D networks and thus, make up the metal oxide nanostructures. A slow and controlled progression of the condensation thereby ensures high crystallinity of the forming oxide material.

In this chapter we describe the chemical mechanisms leading to the formation of M-O-M bridges as basic unit of every metal oxide. It is interesting to note that in spite of the intense research efforts in the area of nonaqueous synthesis of metal oxide nanoparticles, and related to that, of the large number of published methodologies, it is possible to summarize most of the condensation reactions in only seven distinct mechanisms: i) Alkyl halide elimination (in section 5.2), ii) ether elimination (in section 5.3), iii) condensation of carboxylate groups (ester and amide eliminations) formally resulting in elimination of H_2O (in section 5.4), iv) C-C coupling of benzylic alcohols and alkoxide precursors (in section 5.5), v) aldol/ketimine condensation, which may be preceded by solvolysis of the precursor as in the case of acetylacetonates (in section 5.6), vi) oxidation of metal nanoparticles (in section 5.7) and vii) thermal decomposition methods (in section 5.8). Only very few other mechanisms, presented in section 5.9, have been reported so far. Although this list is, in some cases, rather arbitrary, it is a first step towards the rationalization of nonaqueous approaches and should help to develop some general synthesis concepts.

5.2 Alkyl Halide Elimination

The reactions between metal halides and alcohols are doubtlessly among the most utilized methods for the nonaqueous synthesis of metal oxide nanoparticles, mainly due to the broad availability of metal halides and the general advantages of using alcohols as nontoxic solvents. The oxygen of the alcohol molecules is easily coordinated to the metal center of the precursor, which is then followed by an elimination reaction. First systematic investigations on alkyl halide elimination reactions were carried out for metal oxide gels. They involve the condensation between metal alkoxides and metal halides under release of an alkyl halide (Eq. 5.1):

$$\equiv\text{M—Cl} + \text{R—O—M}\equiv \longrightarrow \equiv\text{M—O—M}\equiv + \text{R—Cl} \quad (5.1)$$

The metal alkoxide species is often produced *in situ* by the alcoholysis or the etherolysis of metal halides. In the first case, the metal halide is directly mixed with the alcohol, reacting to the respective metal alkoxo-halide and hydrogen halide HX (Eq. 5.2). Most of the metal halides undergo only partial solvolysis and, depending on the metal center, the number y of the exchanged halides varies strongly [10].

$$\text{MX}_x + y\,\text{ROH} \longrightarrow (\text{RO})_y\text{MX}_{x\text{-}y} + y\,\text{HX} \quad (5.2)$$

$$\equiv\text{M—X} + \text{ROH} \longrightarrow \equiv\text{M—OH} + \text{RX} \quad (5.3)$$

$$\equiv\text{M—OH} + \text{XM}\equiv \longrightarrow \equiv\text{M—O—M}\equiv + \text{HX} \quad (5.4)$$

As a second step, the alkoxide groups formed *in situ* react with unsolvolyzed M-Cl species according to Eq. 5.1. However, alcohols with an R group that is able to stabilize a positive charge, i.e., electron donor substituents like secondary, tertiary carbon, allyl and benzyl groups, prefer to directly eliminate alkyl halides (Eq. 5.3). In this case, a metal hydroxyl species is formed, and therefore the reaction may be adequately termed as a nonhydrolytic hydroxylation process [21, 101]. The condensation between the M-OH and an unsolvolyzed metal halide molecule follows Eq. 5.4. Even though the overall reaction is the same, it is not purely academic to distinguish between these two mechanisms. The intermediate hydroxyl groups are significantly more reactive than alkoxide species, leading to higher rates of gelation in these systems, as it was confirmed by experimental results [21]. For etherolysis, i.e., for the reaction between metal halides and ethers, only one mechanism has been observed (Eq. 5.5), with the second step similar to Eq. 5.1 [2].

$$\equiv\text{M—X} + \text{ROR} \longrightarrow \equiv\text{M—OR} + \text{RX} \quad (5.5)$$

Following these studies on the formation of oxide gels, Trentler et al. published in 1999 the first synthesis of crystalline anatase nanoparticles based on an alkyl halide elimination process [97]. Relatively high temperatures of 300 °C and the presence of trioctylphosphine oxide (TOPO) as surfactant were required to yield phase-pure materials. Those authors proposed that the high-temperature reaction between TiX_4 and $Ti(OR)_4$ essentially follows an SN1 mechanism, with reaction rates strongly decreasing with increased branching of the substituent R, however unaffected by the nature of X. Later, Jun et al. applied the same approach to the shape-controlled synthesis of titiania nanocrystyals [51], while Joo et al. [50] and Tang et al. [96] extended the strategy to the preparation of nanocrystalline zirconia, hafnia and solid solutions thereof by reacting the respective metal chloride and metal isopropoxide in hot TOPO. In the case of zirconia, the organic side products were analyzed. The main species found was *i*PrCl, together with substantial amounts of propene, which formed by dehydrochlorination of *i*PrCl as a consequence of the rather high reaction temperature of 340 °C [50].

halides and alcohols in completely surfactant-free systems (cf. also Chapter 4.1.3) [66, 67, 103]. Even though many mechanistic studies were performed on the formation of metal oxide gels from metal halides and metal alkoxides in the 1990s [101], the occurring reactions are by far not trivial or completely understood. The main expected mechanism in these systems is an alkyl halide elimination as shown in Eq. 5.1. For the reaction of a metal halide in an alcohol the elimination occurs between a halide and a metal alkoxide rather than the alcohol directly, due to previous ligand exchange reactions (Eq. 5.2) [10]. However, this pathway is different for tertiary and benzylic alcohols as they prefer to directly eliminate alkyl halides leading to metal hydroxyl species (Eq. 5.3) [6]. Moreover, the studies of Wang and co-workers showed that in the synthesis of titania from $TiCl_4$ in various primary alcohols, ether formation can also occur, and even become the dominant byproduct observed [103]. Instead of a condensation between a metal alkoxide and a metal halide, as displayed in Eq. 5.1, two *in situ* formed metal alkoxides might also react with each other under formation of a M-O-M bond and an organic ether (instead of an alkyl halide, see also section 5.3). In the original work, the authors assumed that dissolved HCl acts as catalyst for ether formation reactions. In the case of titania nanoparticle synthesized from $TiCl_4$ and benzyl alcohol, alkyl halide formation is reported (Eq. 5.3) as the main pathway due to stabilization of the transient carbo-cation by the mesomeric effect of the phenyl ring [101]. However, addition of small amounts of surface modifiers such as dopamine to the reaction mixture significantly alters the ratio of the reaction products. Whereas in the pure $TiCl_4$ – benzyl alcohol system the final reaction solution contained a molar ratio of 3 benzyl chloride : 1 benzyl ether, the dopamine system in fact contained 1 benzyl chloride : 3 benzyl ether. In all these reaction systems, both benzyl ether and benzyl chloride were present as reaction products, however, varying in amounts depending on the system [28]. In these systems it is therefore difficult to determine which pathway is responsible for the M-O-M formation as the following competing condensation reactions can take place: i) Direct condensation between a metal alkoxide and a metal halide as displayed in Eq. 5.1, ii) condensation involving hydroxyl groups formed as depicted in Eq. 5.4 and, iii) condensation involving ether elimination as described in section 5.3.

Furthermore, one has to keep in mind that the nanocrystals formed in the reaction mixture may even be able to catalyze side reactions such as ether elimination, making it impossible to distinguish between direct products of nanoparticle formation and products of side processes. For example, the reaction between $SnCl_4$ and benzyl alcohol resulted in SnO_2 nanoparticles [4]. During the synthesis the benzyl alcohol solvent, being in 20-fold surplus, is almost completely transformed into dibenzyl ether. As a matter of fact, it was found that tin oxide nanoparticles with surface-adsorbed chloride ions could catalyze the etherification reaction between alcohols by promoting their electrophilicity via a chlorine borrowing mechanism [62]. The presence of small amounts of benzyl chloride led to the conclusion that the formed chloro-alkanes not only were by-products of the reaction, but also acted as intermediates in the formation of the ether, as depicted in Scheme 5.1.

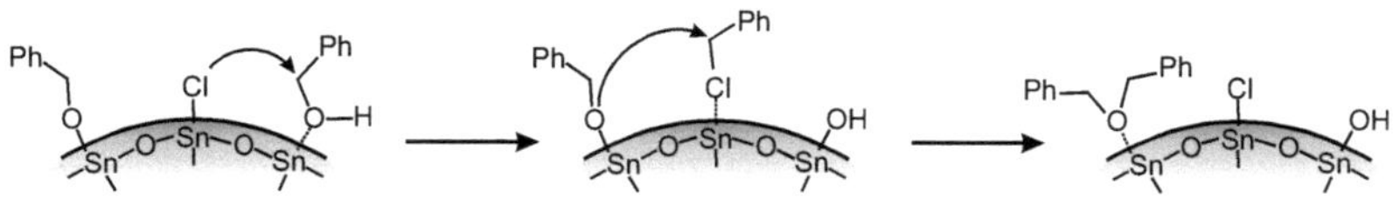

Scheme 5.1. Proposed mechanism for the formation of dibenzyl ether via a chlorine borrowing mechanism catalyzed by SnO_2 nanoparticles

Alkyl halide elimination is also reported for atomic layer deposition (ALD) processes [18]. The first report found in the literature is attributable to Brei et al. [13]. They alternatively reacted $TiCl_4$ and $Si(OEt)_4$ on a silica surface and monitored the silicate growth by IR and studied the by-products of the reaction by mass spectrometry. They observed the formation of chloroethane and thereby proved that an alkyl halide elimination mechanism was involved. Afterward, Ritala et al. [84] showed that the reaction between metal chlorides and alkoxides is generally applicable to the deposition of various metal oxide and silicate thin films (cf. also section 4.3). In contrast to reactions performed in solution, the mechanisms are much more challenging to investigate in the case of ALD. This is based on the fact that the products are in the gas phase and mixed with a large amount of unreacted precursors at each step. Consequently, the quantity of the chemical species available for analysis is very small and highly diluted by the reactants. For this reason, only a limited number of studies were in the position to discuss the possible reaction mechanisms and unfortunately, some of these depositions were investigated under non-ideal conditions, i.e. not in the regime of a self-limiting growth. Moreover, as the deposition temperatures are generally relatively high (300-500 °C) (cf. Table 4.5), the contribution from the metal alkoxide self-decomposition (e.g. β-elimination) cannot be neglected in many studies as it can promote the film growth. A comparison between similar approaches developed in ALD and in liquid phase provides insights and may explain the different behavior observed in terms of reactivity and reaction temperatures. Some studies pointed out that the condensation step involving the alkyl halide elimination needs to be promoted by a metal center acting as Lewis acid [3, 6]. In solution, due to a rapid ligand exchange reaction, different alkoxohalide species are immediately formed [57, 3]. It was proposed that the condensation between $Ti(OiPr)_4$ and $TiCl_4$ leading to the Ti-O-Ti bonds formation follows a concerted mechanism mediated by an alkoxohalide species. In this case, $TiCl_3OiPr$ molecules present in the reaction mixture are the key species for the first condensation reaction converting alkoxo bridges to oxo bridges by the activation of Ti-Cl bonds (Scheme 5.2) [101, 3]. The different behavior

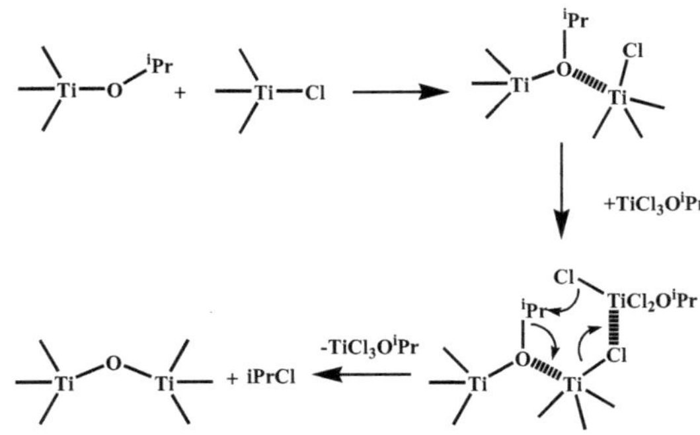

Scheme 5.2 Schematic representation of the condensation between Ti-$OiPr$ and Ti-Cl catalyzed by $TiCl_3OiPr$. Reproduced with permission from Ref. [3]

observed in the ALD of metal oxide via alkyl halide elimination might be explained by these findings. In solution, the condensation reaction does not take place directly between metal alkoxides and metal halides but between intermediate species (e.g. alkoxohalide, oxoalkoxohalide, etc) [57, 3] either because they are more reactive or because they act as catalyst. Thus, the reaction kinetics strictly depends on the concentration of these key species and hence, on the initial chloride/alkoxide concentration [3]. The importance of the presence of these intermediate species is further demonstrated by the observation that an induction period is generally needed to initiate the condensation. Obviously, this reaction can not directly be transposed to ALD. Firstly, because the ratio of reactants is difficult to control and secondly, because the majority of experiments are performed using flow type reactors, i.e. without residence time. Instead, it is expected that the same chemical approach would require higher temperatures in ALD as compared to the liquid phase. Indeed, the deposition processes involving the reaction of a metal alkoxide with a metal halide is reported to efficiently work only for temperatures around 300 °C and above (Table 4.5). Moreover, at these elevated temperatures alkoxy groups at the surface (mainly secondary and tertiary) start to decompose, leaving hydroxyl groups. These groups open an alternative and energetically favorable condensation pathway for the metal halides provided in the subsequent pulse.

5.3 Ether Elimination

Ether elimination involves the formation of the M-O-M bond by the reaction between two metal alkoxides (Eq. 5.6).

$$\equiv M{-}OR + RO{-}M\equiv \longrightarrow \equiv M{-}O{-}M\equiv + ROR \qquad (5.6)$$

Ether elimination was discovered by Bradley et al. during their investigations on the mechanistic aspects of metal alkoxide transformation into oxoalkoxides as intermediates for oxides [7, 8]. Ether elimination reactions are characteristic of the derivatives of multivalent early transition elements such as Mo(VI), W(VI) and Nb(V), where in the presence of pronouncedly basic (alkaline, alkaline earth) alkoxides directly the inorganic molybdates or tungstates are produced (Eq. 5.7) [98, 54].

$$MoO(OEt)_4 + 2\ MOEt \longrightarrow MMoO_4 + 3\ Et_2O \qquad (5.7)$$

$$M = Li, Na;\ Et = CH_3CH_2$$

The formation of the oxo species is analogously possible in reactions of metal alkoxides with aldehydes and ketones [54, 55]. The molybdenum alkoxides $MoO(OCH_3)_4$ and $MoO(OCH_2CH_3)_4$ transform carbonyl compounds like acetaldehyde and acetone into acetals and ketals, respectively, finally yielding the colloidal oxidic compound (Eq. 5.8) [55]. However, the condensation of metal alkoxides in the presence of carbonyl compounds can also proceed along other pathways [40], which is discussed below in section 5.6 (Aldol condensation).

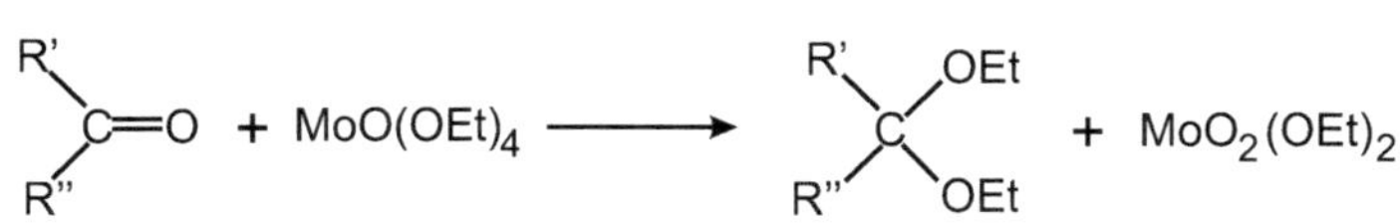

(5.8)

Examples of ether elimination routes to amorphous and nanocrystalline oxides were reported, for example, by Fanelli and Burlew in 1986, who reacted aluminum alkoxides in 2-butanol and obtained amorphous Al_2O_3 in the form of agglomerated particles only 20 nm in size [25]. The glycothermal method, developed by Inoue and co-workers, involved the solvothermal reaction of metal hydroxides and alkoxides in glycols under autogenous pressure at temperatures of 325-450 °C [45, 44, 43, 42]. The formation mechanism of zirconia upon treatment of $Zr(OnPr)_4$ in 1,4-butanediol was proposed as a two-step process. In a first step, 1,4-butanediol binds to the metal center, forming a glycoxide, which in a second step undergoes intramolecular condensation to tetrahydrofuran, which was as a matter of fact identified in the final reaction mixture [42]. This mechanism shows that the dehydration process is catalyzed by the metal itself.

The synthesis of HfO_2 nanoparticles constitutes another example based on the ether elimination process by the solvothermal treatment of hafnium ethoxide with benzyl alcohol [75]. Based on the organic species detected in the final reaction mixture, a two step process involving ligand exchange reactions and ether elimination were proposed. In a first step, a significant amount of ethoxy ligands is exchanged against benzyl alcohol to form ethanol and benzyl alkoxide species (Eq. 5.9). It is known that such ligand exchange processes take place already at room temperature [9, 1]. The alkoxy species undergo condensation to form M-O-M bonds and organic ethers (Eq. 5.10). The amount of dibenzyl ether found in the final reaction mixture was substantially higher than the amount of diethyl ether or benzyl ethyl ether (molar ratio of about 1.0 benzyl ether : 0.3 diethyl ether plus mixed ether), which means that most ethoxy groups are exchanged during the first step in Eq. 5.9. These findings can be explained on the one hand by the surplus of benzyl alcohol in the system, and on the other hand, the better stabilization of transition states due to mesomeric effects of the benzyl group plays an important role, too.

$$\mathrm{Hf(OEt)_4 + x\,C_6H_5CH_2OH \longrightarrow Hf(OEt)_{4-x}(C_6H_5CH_2O)_x + x\,EtOH} \tag{5.9}$$

$$\mathrm{x = 0\text{-}4,\ Et = CH_3CH_2}$$

$$\mathrm{\equiv Hf{-}OR + R{-}O{-}Hf\equiv \longrightarrow \equiv Hf{-}O{-}Hf\equiv + ROR} \tag{5.10}$$

$$\mathrm{R = C_6H_5CH_2\ or\ CH_3CH_2}$$

Also the reaction of titanium alkoxides in benzyl alcohol proceeds via the ether elimination mechanism. Solvothermal treatment of $Ti(O\textit{i}Pr)_4$ in benzyl alcohol at 200 °C for 2 days resulted in nanocrystalline anatase [28]. The final reaction solution contained, in molar ratio, 1.0 benzyl ether : 0.46 isopropyl ethers. The fact that the amount of isopropyl ether and mixed benzyl isopropyl ether is significantly higher as compared to the ethyl ethers found in the HfO_2 synthesis reflects the better stability of secondary carbocations compared to primary ones. The formation of Nb_2O_5 nanoparticles from niobium ethoxide and benzyl alcohol follows an ether elimination pathway, too [28].

A more exotic example of ether elimination was found in the system titanium isopropoxide-benzamide, which yielded at 200 °C anatase nanoparticles of about 5 nm [28]. NMR analysis of the final reaction solution proved the presence of isopropyl ether, N-isopropylbenzamide and isopropyl benzoate. Hence, the main reaction leading to nanoparticle formation is ether elimination. The side reaction, in which a nucleophilic attack of the amide nitrogen on the α-carbon of the isopropoxide occurs, followed by the elimination of an -OH group, also results in the oxide. Such N-alkylation reactions of benzamide with alcohols are unusual, but have been reported to take place under catalysis of transition metals [80].

5.4 Ester and Amide Eliminations

The ester elimination process involves the reaction between metal carboxylates and metal alkoxides (Eq. 5.11).

$$\equiv M{-}O{-}\overset{O}{\overset{\|}{C}}R' + R{-}O{-}M\equiv \longrightarrow \equiv M{-}O{-}M\equiv + RO{-}\overset{O}{\overset{\|}{C}}R' \tag{5.11}$$

This approach has been used for the preparation of alkylsilyloxy derivatives of titanium, zirconium, niobium and tantalum [12, 11] through the reaction of the metal alkoxides $M(OR)_x$ with the silyl esters $R'_3Si(OOCR'')$, yielding the trialkylsilyloxides $M(OSiR'_3)_x$ under elimination of the corresponding ester. Tin(IV) tert-butoxide and trimethylsilyl acetate represent another system, in which the alkoxide group is transferred to the carbonyl ligand of the silyl acetate in nonpolar solvents, similar to acid-catalyzed transesterification [16]. The electropositive and coordinatively unsaturated Sn center acts as a Lewis acid center by coordinating to the carbonyl oxygen atom of the acetate, decreasing the electron density and thus, allowing the nucleophilic attack of the alkoxide.

Compared to ether elimination, the ester elimination process seems to be more versatile, and several examples of nonhydrolytic preparations of (multicomponent) oxide gels have been reported. Anhydrous ZrO_2 and $ZrSiO_4$ gels were synthesized using a direct condensation reaction of metal alkoxides $M(OR)_x$ ($M{=}Zr$; $R{=}C_3H_7$) with metal carboxylates $M'(O_2CR')_y$ ($M'{=}Si, Zr$; $R'{=}CH_3$) releasing a carboxylic acid ester as co-product (Eq. 5.12) [47].

$$M(OR)_x + M'(OOCR')_y \longrightarrow (RO)_{x-1}[M{-}O{-}M'](OOCR')_{y-1} + RCOOR' \tag{5.12}$$

Similarly, $Sn(OAc)_4$ and $Sn(O\mathit{t}Bu)$ yield the tin-oxo cluster $Sn_6(O_6)(O\mathit{t}Bu)_6(OAc)_6$ and *t*BuOAc upon reflux in toluene [14]. The authors proposed several criteria required for successful ester elimination between a metal carboxylate and metal alkoxide: i) a non-chelating carboxylate ligand, ii) a non-coordinating solvent, iii) an electropositive metal alkoxide and, iv) vacant coordination sites on the metal alkoxide center.

The ester elimination route can also be carried out in a modified fashion. Already Pande and Mehrotra investigated the reactions of titanium isopropoxide with acetic anhydride [73]. This approach offers the advantage of a completely hydroxyl-free and thus nonhydrolytic process. It has later been employed to prepare silica gels [27] and TiO_2 nanopowders [37].

Esterification reactions also play an important role in the synthesis of various polyoxometalate and metal oxo clusters [86, 92]. In order to achieve perfectly defined structures, the water required for cluster formation is generated *in situ* through esterification or aldolization reactions in the presence of chelating or bridging organic ligands as stabilizing agents. The preparation involved the use of the corresponding metal isopropoxides in carboxylic acids, giving access to various titanium oxo-isopropoxo clusters [94], a niobium oxo acetate complex [92], and several titanium, zirconium and vanadium oxo clusters [56], just to name a few. Two possible formation mechanisms based on direct esterification (intra coordination sphere mechanism, i.e., water is only virtually formed) or esterification between the slowly exchanged organic acid and the alcohol molecules (water is produced *in situ*) were proposed [94]. Even the direct reaction between metal alkoxides and metal acetates, as shown in Eq. 5.11, in an inert medium produced metal oxide clusters, in which case water could not even theoretically form during the synthesis [15]. Several reviews on titanium oxo clusters were published in the last few years [89, 85, 87].

The elimination of organic esters from alcohol and acid species represent a simple and straightforward approach for the controlled *in situ* formation of water in otherwise anhydrous organic media, and consequently these principles have widely been employed for the synthesis of metal oxide nanoparticles. First experiments were aimed to supply water *in situ* by the reaction of acids with alcohols and the generated water then reacted with the metal oxide precursor, usually metal alkoxides [46], but sometimes also metal halides [102]. However, it is more convenient to directly use the precursors as reactant for esterification. Basically one has the choice between the reaction of metal carboxylates with alcohols, or metal alkoxides with organic acids. The latter approach was particularly successful for the synthesis of various transition metal oxides [49, 112, 59, 22], because the respective alkoxides are available commercially, and certain acids, in particular acids like oleic acid with a long alkyl chain, additionally act as capping and shape-directing agents. On the other hand, metals such as copper and zinc are more convenient in the form of the acetates or carboxylates and therefore, the oxides are synthesized by reaction of the carboxylates with alcohols or diols [39, 48, 107, 19, 5].

Analogous to ester eliminations are amide eliminations based on the reaction of metal oleates with amines, enabling the controlled growth of anisotropic oxides such as titania [111] and zinc oxide nanorods [110].

At this point we have to underline once more that in general it is rather difficult to clearly determine, whether the solvent, surfactant, or the metal organic precursor molecule are the active oxygen donating species. Although the type of precursor and the solvent might point to a specific condensation pathway, it could be that surfactants present in the reaction system lead to another mechanism of M-O-M bond formation. Especially in surfactant-assisted routes, where several types of surfactant are used in the same reaction batch (cf. Table 3.1), for example long-chain amines and carboxylic acids, it could potentially be possible that the surfactants directly react with each other under water formation rather than with the metal containing species.

ALD is an interesting tool for the study of the reaction mechanisms involved in metal oxide formation. in ALD the different reactions are naturally separated due to the alternating introduction of the two reactants. Consequently, ALD can be regarded as a model system suitable for the study of nonaqueous sol-gel reactions. The application of the ester elimination route to ALD was recently reported [81, 82, 104]. It involves the reaction of a metal alkoxide with a carboxylic acid and leads to the formation of metal oxide thin films at low temperature. The proposed reaction mechanism supported by GC-MS and kinetic studies involves an ester elimination via the following two steps. Firstly, during the carboxylic acid pulse, the replacement of the alkoxy ligand at the surface by a carboxylate group takes place under the elimination of alcohol. It is followed, during the metal alkoxide pulse, by a reaction between the carboxylate surface species and the metal alkoxide introduced under the elimination of the ester (Eq. 5.11). In this experiment no water was formed, therefore proving a direct condensation mechanism between alkoxide and acetate species.

5.5 C-C Bond Formation Between Alkoxy Groups

The systematic investigation of C-C bond formation between alcohols started more than 100 years ago by Marcel Guerbet. He studied the reaction of sodium alkoxides with the parent and with other alcohols [35]. He found for example that sodium benzyl alcoholate reacted with ethanol at a temperature of 220 °C to 2-benzylethanol (Eq. 5.13) [35, 36]:

$$C_6H_5CH_2ONa + CH_3CH_2OH \longrightarrow C_6H_5CH_2CH_2CH_2OH + NaOH \tag{5.13}$$

In a broader sense, the Guerbet reaction can be defined as the reaction between two alcohols at high temperatures, generally under catalytic conditions, resulting in a *β*-branched primary alcohol with twice the molecular weight of the reactant alcohol minus a mole of water (Eq. 5.14) [99]. A review on Guerbet chemistry is provided by O'Lenick, Jr. [72].

$$2\ R\text{-}CH_2\text{-}CH_2\text{-}OH \xrightarrow{\text{Catalyst}} \begin{array}{c} R\text{-}CH\text{-}CH_2\text{-}OH \\ | \\ R\text{-}CH_2\text{-}CH_2 \end{array} + H_2O \tag{5.14}$$

It is interesting to note that within Guerbet chemistry, active primary alcohols such as benzyl alcohol or *p*-methoxybenzyl alcohol play a dominant role, and exactly these alcohols are the most popular solvents in the surfactant-free nonaqueous synthesis of metal oxide nanoparticles. It does not come as a surprise either that these alcohols, due to the chemical properties of the phenyl group, are from a chemical point of view responsible for quite fascinating reaction mechanisms for the transformation of metal oxide precursors to oxides. Whereas under "neutral" conditions metal chlorides and alkoxides preferentially undergo the aforementioned alkyl halide and ether elimination mechanisms, in the presence of alkaline species metal alkoxides react with benzyl alcohol via a C-C coupling mechanism, analogous to Guerbet chemistry. This reaction pathway was found during a study on the formation of $BaTiO_3$ nanoparticles prepared by sub-solvothermal reaction of barium metal and titanium isopropoxide in benzyl alcohol [71]. The peculiarity of the reaction system lies in the dissolution of metallic barium in the alcohol prior to addition of the alkoxide $Ti(O\textit{i}Pr)_4$ (Step 1 in Scheme 5.3). It is exactly this additional step that changes the reaction mechanism completely from a common ether elimination process to the C-C coupling pathway. In addition to only negligible quantities of ethers, the major components detected in the final reaction mixture were 4-phenyl-2-butanol and 1,5-diphenyl-3-pentanol. These species were proposed to form via deprotonation of the β-position of the isopropoxide ligand, followed by nucleophilic attack of the benzylic carbon atom [71]. The mechanism, which leads simultaneously to the formation of 4-phenyl-2-butanol and the Ti-O-Ti bond, can be divided into several steps (Step 2 in Scheme 5.3). Upon coordination of benzyl alcohol (shown in red for better illustration) to the Ti atom of the metal oxide precursor, the benzylic carbon atom is activated toward nucleophilic attack (**1**). The dissolution of barium in benzyl alcohol led to the formation of hydrogen gas and benzyl alcoholate (Step 1), which acts as "base" for the deprotonation of the β-carbon atom of the isopropoxy group (**2**). The negative charge is stabilized by interaction with the Ti center (**3**), similar to an agostic bond [24]. This stabilization also explains the deprotonation in β- instead of the α-position. The subsequent nucleophilic attack results in the formation of 4-phenyl-2-butoxide (**4**) and a titanium-bound hydroxyl group. This Ti-OH group promotes further condensation under release of the alcohol (**5**), leading to Ti-O-Ti bridged bonds (**6**) and finally to the oxidic nanoparticles.

The complete absence of acetone and unsaturated species as well as only small amounts of benzaldehyde and other carbonyl compounds in the final $BaTiO_3$ reaction mixture strongly supports this mechanism. Furthermore, this C-C coupling mechanism was not observed when reacting solely $Ti(O\textit{i}Pr)_4$ in benzyl alcohol under otherwise equal experimental conditions. In this case, benzyl ether or isopropyl ethers were formed (cf. Chapter 5.3), as initially expected, yielding crystalline anatase nanoparticles. On the other hand, if alkaline species such as Na(OEt) or K(O*t*Bu) are present, then the C-C coupling mechanism occurred again to a comparable extent as for the $BaTiO_3$ system, however leading to amorphous gels [28]. These results point to the conclusion that obviously "alkaline" conditions are a prerequisite for the occurrence of the C-C coupling mechanism in the Ti system. The coupling reaction is rather insensitive regarding the alkoxide precursor as also $Ti(OEt)_4$, $Ti(O\textit{n}Pr)_4$ and $Ti(O\textit{n}Bu)_4$ could be used as precursors, resulting in the respective coupling products 3-phenylpropanol, 3-phenyl-2-methylpropanol and 2-benzylbutanol in good to almost stoichiometric yields [28]. However, up to now it seems that the C-C coupling mechanism is limited to benzylic carbon atoms and hence to benzylic alcohols as solvents. For the system Ba - $Ti(O\textit{i}Pr)_4$ - 1-phenylethanol, crystalline $BaTiO_3$ nanoparticles were obtained

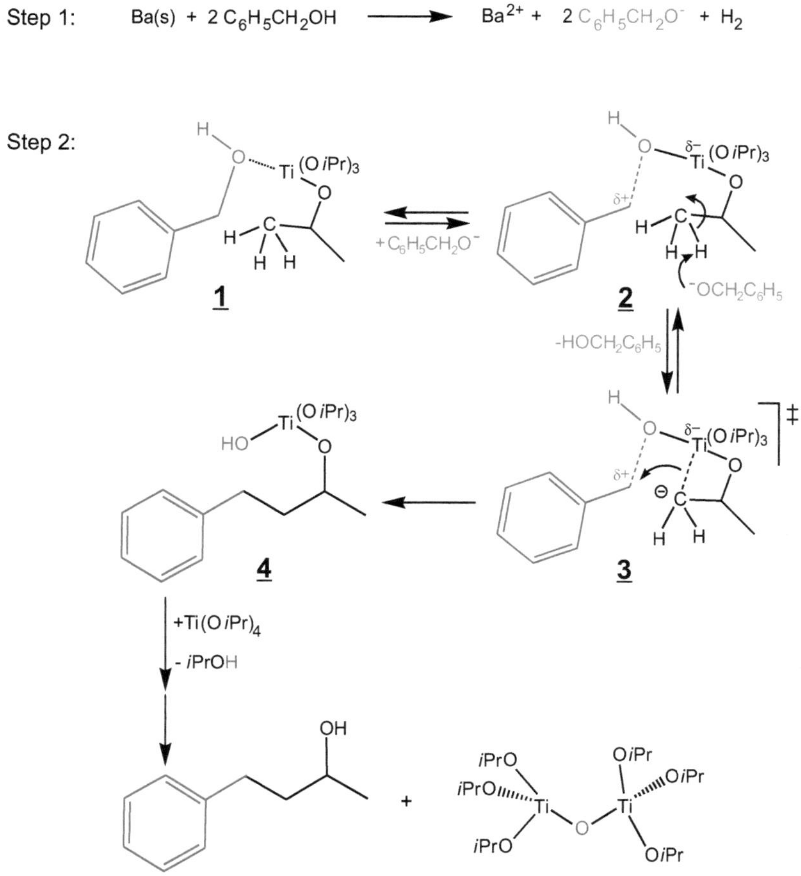

Scheme 5.3. Proposed reaction mechanism for the simultaneous formation of a Ti-O-Ti bridged bond as the basic unit of the $BaTiO_3$ nanoparticles and 4-phenyl-2-butanol

together with 2-phenyl-4-pentanol as the product of the C-C bond formation between an isopropoxy ligand and the solvent. 2-Phenylethanol, with the alcohol function bound to the vicinal instead of the benzylic carbon, does not undergo nucleophilic attack and only amorphous gels were obtained.

The synthesis route to $BaTiO_3$ nanoparticles in benzyl alcohol represents an instructive example to reveal and highlight some basic principles of nonaqueous sol-gel processes. On the one hand, the presence of 4-phenyl-2-butanol as main species and the near absence of organic ethers in the final reaction solution prove that the formation of the metal oxide does not involve an ether elimination process. On the other hand, the retro-synthetical analysis of 4-phenyl-2-butanol provides a mechanistic model on a molecular level, how the oxygen is formally transferred from the solvent benzyl alcohol to the titanium center, finally building up the metal oxide unit. Systematic variation of the synthesis conditions then makes it possible to verify and to gain further information about the parameters that direct the reaction along a specific pathway. All in all, this knowledge helps to understand nanoparticle formation on a chemical basis, which is without any doubts a crucial step toward finding a relationship between the structural, morphological and composition characteristics of the nanoparticles and the initial synthesis system. In the case of $BaTiO_3$, the high crystallinity of the nanoparticles is presumably a direct consequence of the complexity of the organic side reactions, because the slow and controlled provision of the hydroxyl species as condensation agents decreases the reaction rates and facilitates the formation of this ternary oxide in nanocrystalline form.

Analogous to $BaTiO_3$, also the growth of $NaNbO_3$ nanoparticles in benzyl alcohol from metallic sodium and niobium ethoxide proceeds along a C-C bond formation [28, 30]. The use of an ethoxide, instead of an isopropoxide like in the case of $BaTiO_3$, leads to 3-phenylpropanol as condensation product between the ethoxy ligand and benzyl alcohol [28]. However, in comparison to the $BaTiO_3$ system with 4-phenyl-2-butanol as nearly stoichiometric product, the final reaction solution of the $NaNbO_3$ nanoparticles contained more than 20 different organic species, which makes the establishment of the main reaction pathway a difficult task. It is worth to mention again that the reaction between niobium ethoxide and benzyl alcohol (without sodium and also analogous to $BaTiO_3$) proceeds along an ether elimination, evidenced by the presence of benzyl ether and diethyl ether.

Whereas in the case of TiO_2, $BaTiO_3$ and $NaNbO_3$ the presence of a basic species was a prerequisite for C-C bond formation [28, 69], metals with high Lewis acidity, such as many rare earth elements (Y, Ce, Gd, Nd, Sm), can directly catalyze this Guerbet-like reaction [69, 77, 70, 53]. The absence of a formal base could be compensated by the Lewis acidity of these metal complexes, as the formation of an agostic bond (**3** in Scheme 5.3) is expected to be greatly enhanced by the attraction of electrons to these metal centers. The reaction between cerium(III) isopropoxide and benzyl alcohol at 200 °C yielded uniform CeO_2 nanocrystals with diameters of about 3-4 nm and 4-phenyl-2-butanol and 1,5-diphenyl-3-pentanol (result of a second substitution step) as main organic condensation products [70]. Although these mechanisms explain the formation of cerium oxide, i.e., the formation of a Ce-O-Ce bond, they do, however, not give any answers to the question, why CeO_2 is the final composition, although starting from a Ce(III) precursor. Obviously, the cerium was oxidized in the course of the reaction. But in the reaction solution no corresponding reduced species were detected in significant amounts and the oxidation mechanism of cerium remains unclear. In contrast to all the metal oxides discussed up to here in this chapter, in the case of the rare earth metals the reaction between the metal isopropoxides and benzyl alcohol led to lamellar organic-inorganic hybrid nanostructures instead of nanoparticles (cf. Chapter 4.2). The morphological differences between TiO_2, $BaTiO_3$, $NaNbO_3$ and CeO_2 on one, and the rare earth oxides on the other side (in the following discussed for yttrium oxide-benzoate as representative example) are also reflected in small, but crucial variations in the organic reaction pathways.

In contrast to the $BaTiO_3$ system the final reaction solution after formation of the yttrium oxide-benzoate composite contained large amounts of toluene [77]. As a matter of fact, the concentration of toluene was much higher than that of the C-C-coupling product 4-phenyl-2-butanol. In addition, no 4-phenyl-2-butanone was found in the solution, but a significant amount of benzaldehyde. Based on these data, it was proposed that yttrium oxide is able to catalyze, in addition to the C-C bond formation between the isopropoxy ligand and benzyl alcohol, another slightly different reaction, involving two benzyl alcohol molecules (Scheme 5.4). This mechanism is analogous to the disproportionation of benzyl alcohol observed on alumina surfaces at 300 °C [91]. It is an open question whether the reaction is actually catalyzed by just an individual yttrium alkoxide or, more probably, on the surface of the already formed yttrium oxide nanostructures. As both reactions would proceed quite similarly, for simplicity of presentation the former case is discussed. Due to the large excess of benzyl alcohol, it is likely that most of the isopropoxy ligands are exchanged against benzyl alcohol. As another benzyl alcohol molecule coordinates to the metal center (**1**), a hydride transfer can take place between the CH_2 group of the coordinated benzyl alcohol ligand and the CH_2 group of the second ligand (**2**). The next step (**3**) involves the release of toluene and benzaldehyde. At the same time a hydroxy group is formed on the yttrium species, promoting the condensation with another yttrium isopropoxide. As benzaldehyde is found in much lower proportion than

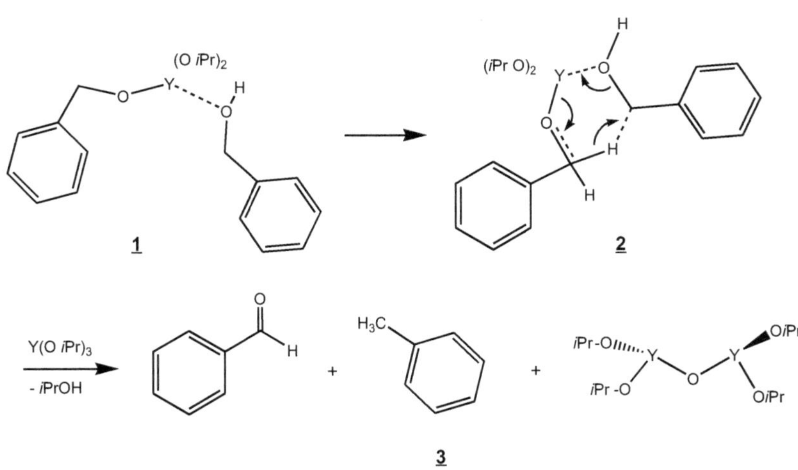

Scheme 5.4. Secondary hydride-transfer reaction leading to the disproportionation of benzyl alcohol to toluene and benzaldehyde

toluene (about 1:4), but the nanocomposite contains larger amounts of benzoate molecules, it was assumed that the yttrium oxide nanostructure was catalyzing another hydride-transfer reaction. The formation of benzoic acid via a Cannizzaro-like reaction on various metal oxides is known, and even the simultaneous formation of metal-coordinated benzoic acid and toluene from benzyl alcohol and benzaldehyde has been suggested [38]. The proposed mechanism is displayed in Scheme 5.5. First, benzaldehyde coordinates to an yttrium center, producing a highly electrophilic carbon atom (**1**). In a next step, the oxygen atom of a benzyl alcohol molecule, which is coordinated to a vicinal yttrium center, attacks the electrophilic carbon. Simultaneously, hydride transfer takes place to form toluene (**2**). As the resulting benzoate is coordinated to the yttrium oxide surface (**3**), it blocks further 3D growth of the nanocrystal. IR and Raman spectra of the yttria-benzoate hybrid material pointed to a bonding of two yttrium atoms to one carboxylate group, as proposed in Scheme 5.5 [77]. Both reaction and structure formation are therefore self-confining. The generated surface-bound acid stops further structural growth and defines the geometry of a lamellar nanostructure of alternating yttria-benzoate layers, whereas no further benzoate is generated when the surface is saturated. This self-confinement also speaks for the fact that the presence of a minimal small-sized yttria surface is a prerequisite for the side reaction to form benzoate molecules. This proposed mechanism was verified by reacting the yttrium alkoxide directly with benzaldehyde instead of benzyl alcohol. To a great extent, benzyl benzoate $C_6H_5CH_2OCOC_6H_5$, which is the expected product of two benzaldehyde molecules reacting via the proposed mechanism, was found in the reaction solution. This proves the occurrence of hydride transfer reactions under these reaction conditions.

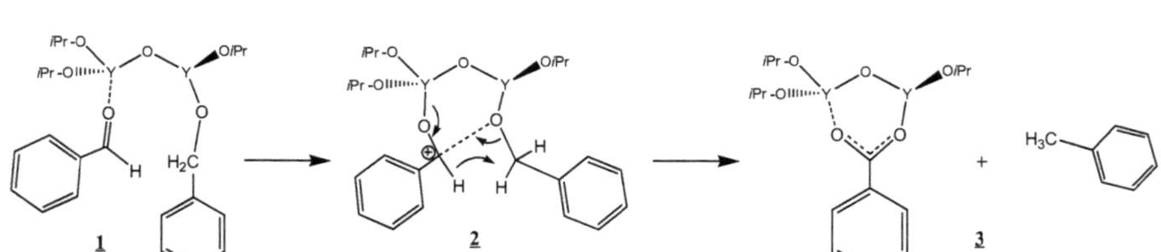

Scheme 5.5. Proposed formation mechanism of the yttria-benzoate nanohybrid

5.6 Aldol/Ketimine Condensation

Most of the sol-gel procedures to metal oxides rely (aside from water) on the use of alcohols, amines or acids as solvent. However, other organic compounds such as ketones, aldehydes, esters and nitriles also constitute attractive reaction media. Especially ketones represent a viable alternative to alcohols offering a less reductive synthesis environment. In the case of ketones and aldehydes, the release of oxygen generally involves aldol condensation reactions under elimination of water and formation of α,β-unsaturated carbonyl compounds (Eq. 5.15). Such aldol elimination processes are known to be catalyzed by metal complexes [83], which are anyway present in the system as the metal oxide precursors. The eliminated water is then consumed for the formation of the oxide, thus driving the equilibrium to the side of the condensation products.

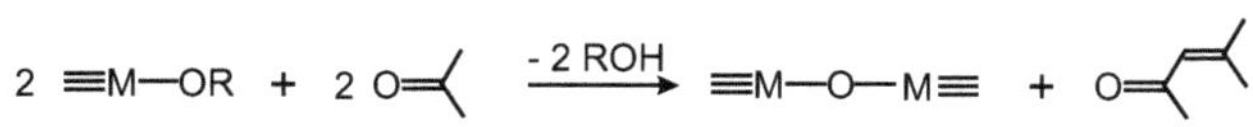

(5.15)

Aldol condensation processes can take place at moderate temperatures, as demonstrated by Goel et al. [34]. They reacted $Zn[OC(CH_2CH_3)_3]_2$ with acetone in benzene or toluene at room temperature, leading to transparent gels that transformed to precipitates consisting of hydrated zincite nanocrystals upon aging. The reaction pathway involves the deprotonation of acetone and coordination to Zn to form an enolate complex. In a second step, the enolate ligand nucleophilically attacks another acetone species via aldol condensation to yield Zn-OH and mesityl oxide as condensation products. In an analogous approach, titanium oxo clusters were prepared by the reaction of titanium isopropoxide with ketones such as acetone, acetylacetone and diacetone alcohol at room temperature [93]. Organic condensation products detected in the acetone system were mesityl oxide and the tridentate ligand 2,6-dimethylhept-3-en-2,4,6-triol ($O_3C_9H_{18}$). Based on these findings, the authors proposed a reaction pathway as depicted in Scheme 5.6 and which is quite similar to the mechanism reported by Goel et al. for the formation of zincite [34]. After the coordination of acetone to the titanium center, deprotonation leads to the enolate complex under concurrent release of isopropanol (**1**). The enolate ligand nucleophilically attacks a second acetone molecule to form a C-C bond (**2**), and the aldolate ligand is coordinated to the Ti center (**3**). This titanium diacetone alcoholate complex (**3**) can either react to a Ti-OH species under release of mesityl oxide and further proceed to the formation of a Ti-O-Ti bond (**4**), or undergo further condensation reaction with a third acetone molecule [93]. In the latter process, complex (**5**) is transformed into the tridentate ligand (**6**), which is an excellent stabilizing agent. If the reaction is performed at room temperature, well-defined titanium oxo clusters were obtained [93], whereas the solvothermal treatment of $Ti(O\textit{i}Pr)_4$ in ketones and aldehydes extended the size of the products beyond clusters and lead to the formation of nanocrystalline anatase in the size range of 7 to 20 nm, depending on the solvent [29]. In this case, even higher condensation products of acetone such as phorone and mesitylene were observed.

The reaction of titanium isopropoxide with higher aliphatic ketones like 2-butanone and 3-pentanone as solvent leads additionally to acetone and

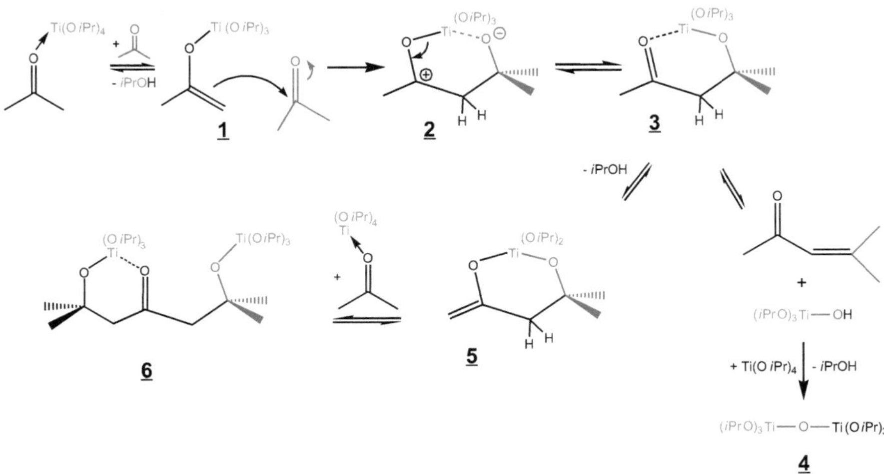

Scheme 5.6. Proposed formation mechanism of TiO_2 in acetone via aldol condensation

2-butanol or 3-pentanol. These species are the result of a reduction-oxidation mechanism between the isopropoxide ligand and the ketone solvent, mediated by the Ti center, and analogous to a Meerwein-Ponndorf-Verley reaction found upon heating of aluminum alkoxides in acetone [64, 100, 79]. This pathway has also been observed as side reaction in the formation of $BaTiO_3$ in cyclohexanone [33], where via a hydride transfer from the alkoxide ligand to a coordinated solvent molecule, the ligand is oxidized to the respective ketone whilst the solvent is reduced to an alkoxy group. The mechanism for the hydride transfer reaction is presented in Scheme 5.7. In the first step the solvent coordinates to the Ti center (**1**), followed by the hydrogen transfer from the α-position of the isopropoxide ligand to the ketone via a hydride transfer in a six-membered cyclic transition state (**2**) [88]. The reduced ketone solvent is then bound to the Ti as an alkoxide, and acetone is released (**3**). Upon condensation, also the reduction product is released as secondary alcohol. One has to keep in mind that this side reaction does, by itself, not contribute to the oxide formation and therefore, aldol condensations as shown in Scheme 5.6 represent the essential pathway for nanoparticle formation.

Also the preparation of $BaTiO_3$ nanocrystals 6-15 nm in size was achieved at room temperature by the long-time stirring of solutions of a pre-synthesized mixed barium-titanium oxoalkoxide in acetone [33]. If the reaction is performed at higher temperatures, the process is accelerated. For example, solvothermal treatment of a reaction mixture consisting of metallic barium dissolved in various ketones such as acetone or acetophenone and $Ti(O\mathit{i}Pr)_4$ resulted in $BaTiO_3$ nanoparticles with crystallite sizes in the range of 10-15 nm [68]. NMR analysis of the final reaction solution gave evidence for

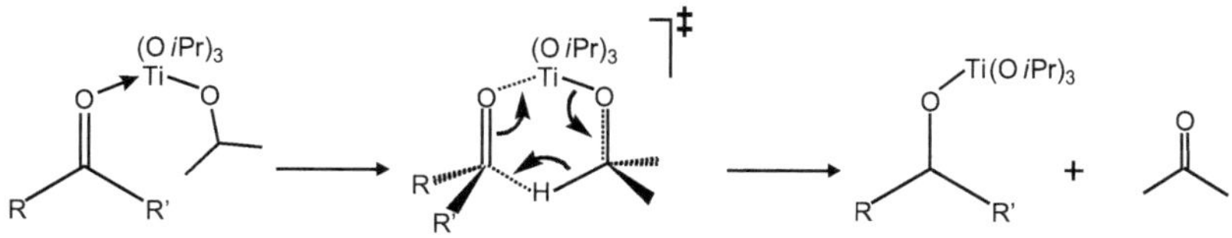

Scheme 5.7. Side reaction observed during the formation of TiO_2 nanoparticles in higher aliphatic and aromatic ketones, analogous to the Meerwein-Ponndorf-Verley reaction

the presence of isopropanol, acetone, 1-phenylethanol and 1,3-diphenyl-2-buten-1-one, in addition to the solvent acetophenone. Also mesityl oxide and 1-phenyl-3-methyl-2-buten-1-one were found in traces. The presence of acetone and 1-phenylethanol can be explained by a redox process analogous to the Meerwein-Ponndorf-Verley reaction in Scheme 5.7. Based on the fact that 1,3-diphenyl-2-buten-1-one is the main reaction product, the reaction mechanism is proposed as illustrated in Scheme 5.8. In a first step, metallic barium dissolves in acetophenone under formation of hydrogen gas, which is visible by the appearance of gas bubbles in the reaction mixture. The resulting barium enolate (**1**) induces a red-brown color of the solution. Upon addition of titanium isopropoxide (Step 2), the nucleophilic enolate coordinates readily to the Ti center (**2**) and is thus stabilized. During thermal treatment in the autoclave, it attacks a second acetophenone (**3**) similar to an aldol coupling mechanism, mediated by the Ti center. Formal elimination of water (**4**) leads to the formation of 1,3-diphenyl-2-buten-1-one under concurrent release of isopropanol (**5**), which is the key step in this mechanism, because it provides the oxygen necessary for the formation of crystalline $BaTiO_3$. The Ti-OH groups, resulting from this last reaction step, can react with another titanium isopropoxide under release of isopropanol to form Ti-O-Ti bonds.

Aldol condensation reactions were also found as a part of the formation mechanism on the way to metal oxides starting from metal acetylacetonates and benzyl alcohol [78] or benzylamine [76]. The solvothermal reaction of metal acetylacetonates with these solvents yielded nanocrystalline Fe_3O_4, In_2O_3, Ga_2O_3 and ZnO. According to the organic species detected in the final synthesis liquid the reaction started with a solvolysis of the acetylacetonate species, followed by aldol or ketimine condensation reactions. The mechanism for the reaction of $Fe(acac)_3$ in benzyl alcohol is depicted in Scheme 5.9. After the nucleophilic attack of the benzyl alcohol on one of the carbonyl groups of the acetylacetonate ligand (**1**), alcoholysis leads to benzyl acetate and an enolate ligand (**2**). In a next step, benzyl alcohol coordinates to the Fe center, releasing benzyl acetate in a ligand exchange reaction. The enolate attacks the coordinated benzyl alkoxide (**3**, **4**) and the subsequent

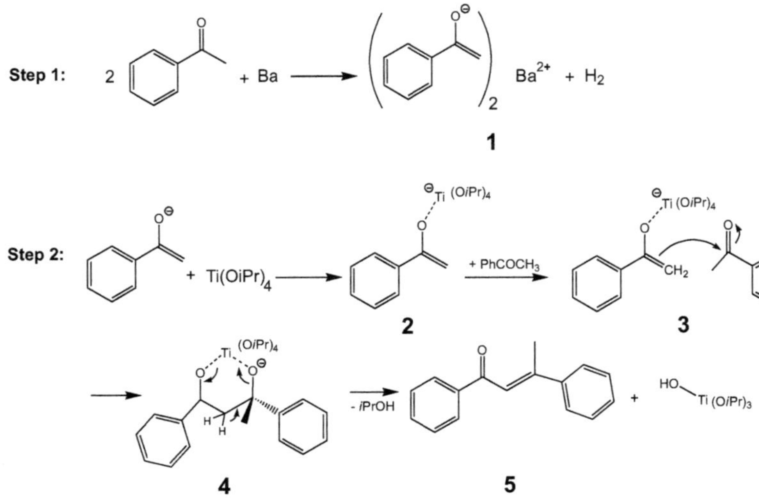

Scheme 5.8. Proposed main reaction mechanism leading to the simultaneous formation of 1,3-diphenyl-2-buten-1-one and Ti-OH groups, which then further condense to form $BaTiO_3$ nanoparticles

release of 4-phenyl-2-butanone (**5**) leads to the formation of a Fe-OH species, representing the starting point of nanoparticle formation.

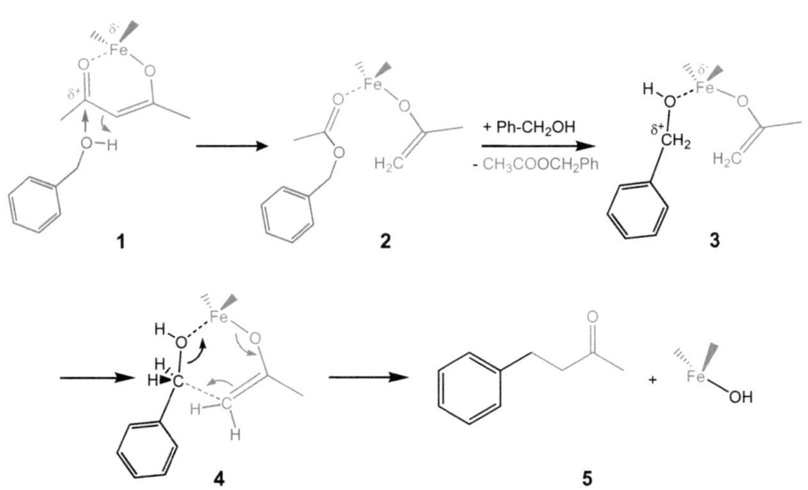

Scheme 5.9. Main reaction occurring upon solvothermal treatment of $Fe(acac)_3$ in benzyl alcohol involving solvolysis of the acetylacetonate, followed by condensation reactions

The formation of magnetite Fe_3O_4 nanoparticles starting from Fe(III) is only possible upon partial reduction of the iron center to Fe(II). It was proposed that dehydrogenative oxidation of 4-phenyl-2-butanone (**5**) to 4-phenyl-3-buten-2-one was responsible for reducing the iron centers, and as a matter of fact, 4-phenyl-3-buten-2-one was detected in the final reaction solution by GC-MS in considerable amount [69].
The reaction of $Fe(acac)_3$ in benzylamine proceeds analogously to Scheme 5.9, with the second step of the reaction however being a ketimine condensation (Scheme 5.10), where the enolate ligand plays the role of an electrophile (in contrast to the reaction in benzyl alcohol, where the enolate ligand was the nucleophile). Due to the higher Lewis basicity of benzylamine, the solvolysis is enhanced (**1**), resulting in N-benzylacetamide and an enolate species (**2**), which undergoes the ketimine condensation (**2**, **3**) to form N-isopropylidene-benzylamine and an Fe-OH species (**4**).

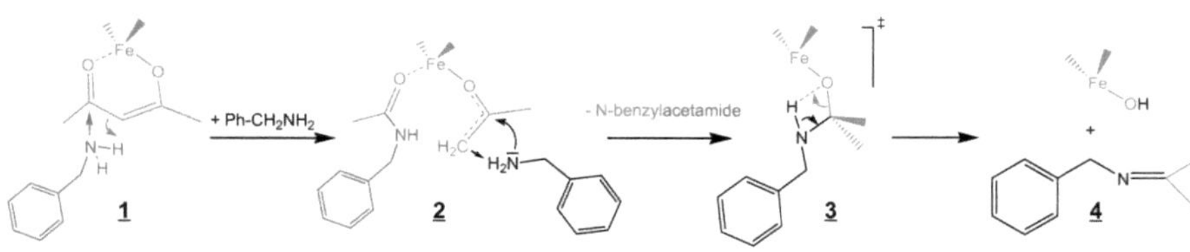

Scheme 5.10. Main pathway proposed for the reaction of $Fe(acac)_3$ with benzylamine leading to the formation of the iron hydroxyl species and N-isopropylidenebenzylamine

Titanium oxide nanoparticles have also been synthesized using aldehydes as oxygen-supplying agents [29]. We will discuss the titanium isopropoxide-benzaldehyde system as representative example. In addition to benzaldehyde, substantial amounts of benzyl alcohol, a reduction product of benzaldehyde, and of benzyl benzoate, a disproportionation product of benzaldehyde, were detected. Interestingly, a significant amount of 4-phenyl-2-butanone was found too, being the analog to 4-phenyl-2-butanol, which has been found in large amounts for the synthesis of $BaTiO_3$ in benzyl alcohol (cf. Section 5.5). In that case, the formation of 4-phenyl-2-butanol was proposed to proceed via a C-C coupling reaction from $Ti(O\textit{i}Pr)_4$ and benzyl alcohol under basic conditions during the oxide formation. Therefore, in the benzaldehyde system, one can assume intermediate formation of 4-phenyl-2-butanol, which is subsequently oxidized by the aldehyde to form the ketone. On the other hand, the aldol addition product of acetone to benzaldehyde, namely 4-phenyl-3-buten-2-one, could also be identified. Based on these findings, a relatively complicated reaction sequence leading to the formation of TiO_2 was proposed (Scheme 5.11). In a first step (Step 1), a Meerwein-Ponndorf-Verley-like reaction of benzaldehyde with $Ti(O\textit{i}Pr)_4$ leads to acetone and benzyl alcohol (1). Then, there are two possible routes (Step 2, Pathway 1 and 2). The alcohol can react (Pathway 1) with another isopropoxy ligand via the aforementioned C-C coupling reaction (2). Due to the surplus of benzaldehyde as oxidizing agent, the resulting 4-phenyl-2-butanol is readily oxidized to 4-phenyl-2-butanone (3). For the second pathway (Pathway 2), aldol addition of acetone to the aldehyde (4,5) was proposed. In the case of aromatic aldehydes, Pathway 1 is favored as the intermediate carbocation is stabilized, whereas in aliphatic aldehydes aldol coupling reactions as proposed in Pathway 2 are the more probable formation mechanism for TiO_2.

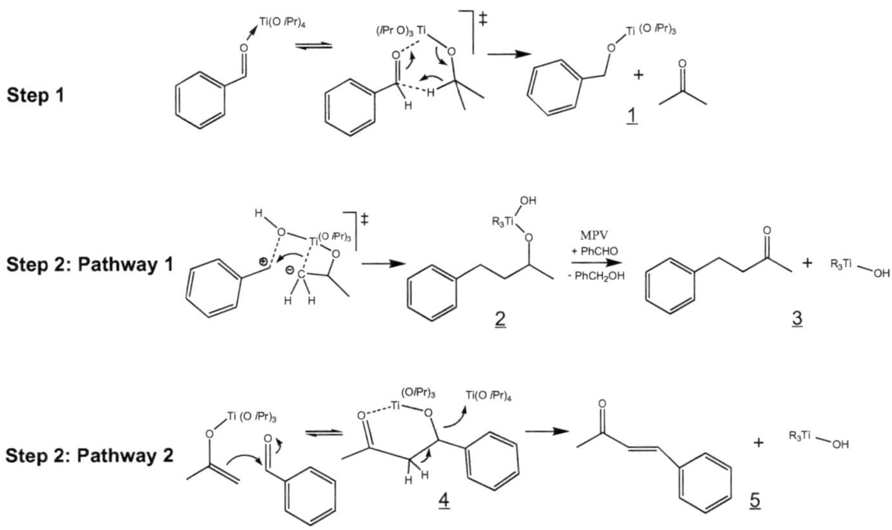

Scheme 5.11. Reaction pathways identified for the synthesis of TiO_2 nanoparticles in benzaldehyde (MPV = Meerwein-Ponndorf-Verley)

The interested reader is additionally referred to Ref. [28], which includes more peculiar examples and side reactions found in the nonaqueous and surfactant-free synthesis of metal oxide nanoparticles, together with a lot of experimental details.

5.7 Oxidation of Metals

If the metals are easy to achieve on the nanoscale, the conversion of the metal nanoparticles into the oxides via an oxidation process constitutes another alternative route, as reported for FeO_x, NiO, ZnO or CuO [41, 52, 108, 23, 74]. Although mild organic oxidants such as trimethylamine oxide can be used, in the simplest case even exposure to air or stirring in organic solvents in contact with air for a longer period of time is sufficient to complete oxidation [108, 74, 26]. This approach takes advantage of the good size and shape control as provided by some preparation methods for nearly monodisperse metal nanoparticles. If carefully performed, no alteration of particle size and morphology occurs during transformation into the oxide. This two step process, synthesis of metal nanoparticles and subsequent oxidation to the oxide, can be simplified in a way that the oxidizing agent is either directly added to the initial synthesis mixture, or the synthesis is performed under aerobic conditions [41, 106, 17, 58, 105]. Similarly, core-shell nanoparticles consisting of a metal core and a shell of the respective metal oxide are accessible [90], as further discussed in Chapter 3.8.

5.8 Other Mechanisms

In addition to these more generally applicable synthesis methodologies, more specialized, but nonetheless interesting strategies were reported in literature. Thermal decomposition reactions in a narrower sense, where elimination of alkenes from alkoxides leaves hydroxyl groups that can induce further condensation, have already been proposed by Mazdiyasni et al. [63] and shall not be discussed further here. Another example is the reaction of bis(cyclooctatetraene)titanium, $Ti(COT)_2$, with dimethyl sulfoxide (DMSO) in o-dichlorobenzene at room temperature [95]. The oxygen is formally supplied by DMSO, which is reduced by $Ti(COT)_2$ to the disulfide (Eq. 5.16). It is a remarkable feature of this route that, in contrast to the other low-temperature methods discussed before, at no stage there is a formation of hydroxyl groups or water.

$$Ti(COT)_2 + 2\,(CH_3)_2SO \longrightarrow TiO_2 + 2\,S(CH_3)_2 + 2\,COT \qquad (5.16)$$

COT = 

The reaction of titanium isopropoxide with benzylamine yielded anatase nanoplatelets stacked together in a lamellar fashion with a small organic layer in between (cf. Chapter 4.2.4) [32]. In GC-MS analysis of the final reaction solution benzylamine was identified as the main component, in addition to dibenzylamine, N-benzylidenebenzylamine, N-isopropylbenzylamine, isopropanol and toluene in smaller quantities. The main reaction occurring in the $Ti(OiPr)_4$-benzylamine system thus is a condensation between two benzylamine molecules, as illustrated in Scheme 5.12, Pathway 1. It can be expected that the condensation is greatly catalyzed by the Ti centers, as the high affinity of amino groups to Ti alkoxides is well-known since the studies of Cook in the 1950s [20]. Ammonia, the other product formed in this pathway, is obviously not capable of inducing bridging to another metal center. Instead of the nitride, the thermodynamically much more stable oxide is formed (see below), which implies that ammonia must be released during the further course of the condensation. As a matter of fact, large quantities of ammonia were observed in the reaction system, escaping upon opening the autoclave. The formation of the anatase nanoplatelets hence must be induced via a different side reaction. Quantitative analysis of the products detected by GC-MS indicated that half of the isopropoxy groups must have been converted to volatile compounds during the synthesis. Based on these data, an elimination of isopropoxy ligands of the titanium alkoxide as the main mechanism for the anatase formation was proposed (Scheme 5.12, Pathway 2), leading to gaseous propene as by-product and Ti-OH, which then initiated the formation of Ti-O-Ti units. Through further exchange reactions, also N-isopropylbenzylamine is formed. This mechanism proves the high affinity of the amine nitrogen towards the metal alkoxide, at the same time underlining its inability to form bridges between the metal centers. The same phenomenon has been observed for the reaction of metal acetylacetonates in benzylamine [76], and therefore, one can conclude that the synthesis of metal nitrides in amine solvents cannot be achieved under standard sol-gel conditions starting from metal organic precursors.

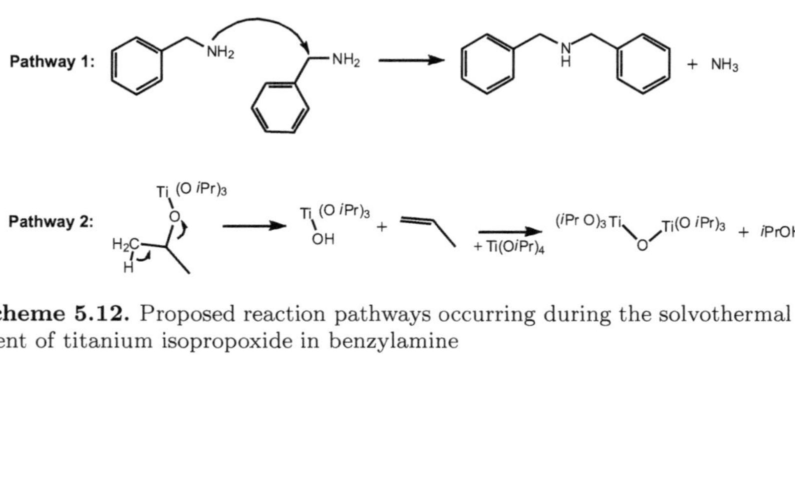

Scheme 5.12. Proposed reaction pathways occurring during the solvothermal treatment of titanium isopropoxide in benzylamine

5.9 Discussion

The chemical reaction mechanisms presented in this chapter are a result of retro-synthetical analysis based on the composition of the final reaction mixture, without any experimental evidence from *in situ* measurements. In this context one has to keep in mind that the nonaqueous synthesis is often performed under quite harsh conditions in comparison to common reaction approaches in organic chemistry. The metal-containing species present in the reaction mixture may act as catalyst not only for their own condensation into an oxide network, but for all sorts of organic reactions. In some cases, the oxygen-supplying reaction pathway is practically superimposed with another condensation reaction. The example of SnO_2 nanocrystal formation in benzyl alcohol, resulting in complete condensation of the alcohol solvent into benzyl ether was already discussed before in Chapter 5.2. Ether formation was also observed to be catalyzed during the formation of $InNbO_4$ nanocrystals in benzyl alcohol, and interestingly even led to an aqueous phase that separated from the main organic reaction phase and was by far exceeding the amount of water actually required for the transformation of the precursors into the metal oxide [109]. In both of these cases, it is not easily possible to separate the organic mechanism for the supply of oxygen/water for the oxide formation from the catalyzed condensation reaction proceeding to a far greater extent.

The investigation and classification of these reaction mechanisms represents an important step in metal oxide nanoparticle synthesis, providing information about the oxygen source for metal oxide formation, the chemical mechanisms involved in the transfer of the oxygen from the source to the metal center, and about the formation of various organic species that potentially act as coordinating, and thus, size- and shape-controlling agents [65, 31]. Whereas the reaction pathways offer answers concerning the relationship between a particular synthesis system and the final composition of the metal oxide, this is hardly the case for the crystallite morphology. It is nearly unpredictable, which crystallite size and shape can be obtained from the reaction of a particular metal oxide precursor with a specific organic solvent. Although the organic species strongly influence the structural, compositional, and morphological characteristics of the inorganic product and thus offer a versatile tool for tailoring particle size, shape, composition, and surface properties, this is only possible on an empirical basis [65, 31]. Nevertheless, it is now widely recognized that efforts to perform mechanistic studies for newly developed approaches to oxide nanoparticles are worthwhile, leading to surprising results. For example, when Li et al. presented a novel nonaqueous method to Fe_3O_4 nanocrystals in 2-pyrrolidone, the transformation of the precursor $Fe(acac)_3$ to Fe_3O_4 was simply termed as a "thermal decomposition", not giving any indication about why one third of the iron ions are reduced to Fe(II) required for the formation of Fe_3O_4 [60]. Upon reporting a slightly altered protocol in their following publication, however, they also included mechanistic studies based on photoelectric spectroscopy analysis of the gaseous reaction products. These investigations revealed that in fact 2-pyrrolidone is unstable under the employed reflux conditions, releasing carbon monoxide, which leads to a partial reduction of Fe(III) to Fe(II) [61]. This example highlights again the role of the organics in the reaction system. Obviously, the solvent is crucial not only with respect to the stabilization of nanoparticles, leading to control over the size and shape, but also towards the intrinsic composition of the material.

References

1. Andrianainarivelo, M., Corriu, R., Leclercq, D., Mutin, P.H., Vioux, A.: Mixed oxides SiO_2-ZrO_2 and SiO_2-TiO_2 by a non-hydrolytic sol-gel route. J. Mater. Chem. **6**, 1665–1671 (1996)
2. Arnal, P., Corriu, R.J.P., Leclercq, D., Mutin, P.H., Vioux, A.: Preparation of transition metal oxides by a nonhydrolytic sol-gel process. Mater. Res. Soc. Symp. Proc. **346**, 339–344 (1994)
3. Arnal, P., Corriu, R.J.P., Leclercq, D., Mutin, P.H., Vioux, A.: A solution chemistry study of nonhydrolytic sol-gel routes to titania. Chem. Mater. **9**, 694–698 (1997)
4. Ba, J., Polleux, J., Antonietti, M., Niederberger, M.: Nonaqueous synthesis of tin oxide nanocrystals and their assembly into ordered porous mesostructures. Adv. Mater. **17**, 2509–2512 (2005)
5. Bilecka, I., Djerdj, I., Niederberger, M.: One-minute synthesis of crystalline binary and ternary metal oxide nanoparticles. Chem. Commun. pp. 886–888 (2008)
6. Bourget, L., Corriu, R.J.P., Leclercq, D., Mutin, P.H., Vioux, A.: Non-hydrolytic sol-gel routes to silica. J. Non-Cryst. Solids **242**, 81–91 (1998)
7. Bradley, D.C., Chakravarti, B.N., Wardlaw, W.J.: Normal alkoxides of quinquevalent niobium. J. Chem. Soc. pp. 2381–2384 (1956)
8. Bradley, D.C., Chakravarti, B.N., Wardlaw, W.J.: Structural chemistry of the alkoxides. Part VIII. Isomeric butoxides and pentyloxides of niobium. J. Chem. Soc. pp. 4439–4442 (1956)
9. Bradley, D.C., Hancock, D.C., Wardlaw, W.: Titanium chloride alkoxides. J. Chem. Soc. pp. 2773–2778 (1952)
10. Bradley, D.C., Mehrotra, R.C., Rothwell, I.P., Singh, A.: Alkoxo and aryloxo derivatives of metals. Academic Press: London (2001)
11. Bradley, D.C., Thomas, I.M.: A new method for the preparation of trialkylsiloxy-derivatives of metals. Chem. and Ind. **38**, 1231–1232 (1958)
12. Bradley, D.C., Thomas, I.M.: Organosilyloxy-derivatives of metals. Part I. Alkylsilyloxy derivatives of Titanium, Zirconium, Niobium and Tantalum. J. Chem. Soc. pp. 3404–3411 (1959)
13. Brei, V.V., Kaspersky, V.A., Gulyanitskaya, N.U.: Synthesis and study of boron phosphate and titanium silicate compounds on silica surface. Reaction Kinetics and Catalysis Letters **50**, 415–421 (1993)
14. Caruso, J., Hampden-Smith, M.J.: Ester elimination: A general solvent dependent non-hydrolytic route to metal and mixed-metal oxides. J. Sol-Gel Sci. Technol. **8**, 35–39 (1997)
15. Caruso, J., Hampden-Smith, M.J., Duesler, E.N.: Solvent dependent ester elimination reactions in the preparation of mixed-metal oxo clusters: The synthesis of $PbSn_2(\mu_3\text{-}O)(OBut)_4(OAc)_4$. J. Chem. Soc., Chem. Commun. pp. 1041–1042 (1995)
16. Caruso, J., Roger, C., Schwertfeger, F., Hampden-Smith, M.J., Rheingold, A.L., Yap, G.: Solvent-dependent ester elimination and ligand exchange reactions between trimethylsilyl acetate and tin(IV) tetra-tert-butoxide. Inorg. Chem. **34**, 449–453 (1995)
17. Cheon, J., Kang, N.J., Lee, S.M., Lee, J.H., Yoon, J.H., Oh, S.J.: Shape evolution of single-crystalline iron oxide nanocrystals. J. Am. Chem. Soc. **126**, 1950–1951 (2004)
18. Clavel, G., Rauwel, E., Willinger, M.G., Pinna, N.: Nonaqueous sol-gel routes applied to atomic layer deposition of oxides. J. Mater. Chem **19**, 454–462 (2009)
19. Clavel, G., Willinger, M.G., Zitoun, D., Pinna, N.: Solvent dependent shape and magnetic properties of doped ZnO nanostructures. Adv. Funct. Mater. **17**, 3159–3169 (2007)

20. Cook, C.M.: Interaction of tetraisopropyl titanate and monoethylamine in n-decane solution. J. Am. Chem. Soc. **81**, 3828–3830 (1959)
21. Corriu, R.J.P., Leclercq, D., Lefevre, P., Mutin, P.H., Vioux, A.: Preparation of monolithic gels from silicon halides by a non-hydrolytic sol-gel process. J. Non-Cryst. Solids **146**, 301–303 (1992)
22. Cozzoli, P.D., Kornowski, A., Weller, H.: Low-temperature synthesis of soluble and processable organic-capped anatase TiO_2 nanorods. J. Am. Chem. Soc. **125**, 14,539–14,548 (2003)
23. Cozzoli, P.D., Kornowski, A., Weller, H.: Colloidal synthesis of organic-capped ZnO nanocrystals via a sequential reduction-oxidation reaction. J. Phys. Chem. B **109**, 2638–2644 (2005)
24. Dawoodi Z., G.M.M.V.P.K.: Evidence for a direct bonding interaction between Ti and a β-C-H moiety in a titanium-ethyl compound. J. Chem. Soc. Chem. Commun. pp. 802–803 (1982)
25. Fanelli, A.J., Burlew, J.V.: Preparation of fine alumina powder in alcohol. J. Am. Ceram. Soc. **69**, C174–C175 (1986)
26. Feng, X., Bai, Y.J., Lu, B., Zhao, Y.R., Yang, J., Chi, J.R.: A novel reduction/oxidation synthetic route to cubic zirconia nanocrystallite. J. Cryst. Growth **262**, 420–423 (2004)
27. Fujiwara, M., Wessel, H., Park, H.S., Roesky, H.W.: A sol-gel method using tetraethoxysilane and acetic anhydride: Immobilization of cubic μ-oxo si-ti complex in a silica matrix. Chem. Mater. **14**, 4975–4981 (2002)
28. Garnweitner, G.: Nonaqueous synthesis of transition-metal oxide nanoparticles and their formation mechanism. Ph.D. thesis, University of Potsdam (2005)
29. Garnweitner, G., Antonietti, M., Niederberger, M.: Nonaqueous synthesis of crystalline anatase nanoparticles in simple ketones and aldehydes as oxygen-supplying agents. Chem. Commun. pp. 397–399 (2005)
30. Garnweitner, G., Niederberger, M.: Nonaqueous and surfactant-free synthesis routes to metal oxide nanoparticles. J. Am. Ceram. Soc. **89**, 1801–1808 (2006)
31. Garnweitner, G., Niederberger, M.: Organic chemistry in inorganic nanomaterials synthesis. J. Mater. Chem. **18**, 1171–1182 (2008)
32. Garnweitner, G., Tsedev, N., Dierke, H., Niederberger, M.: Benzylamines as versatile agents for the one-pot synthesis and highly ordered stacking of anatase nanoplatelets. Eur. J. Inorg. Chem. pp. 890–895 (2008)
33. Gaskins, B.C., Lannutti, J.J.: Room temperature perovskite production from bimetallic alkoxides by ketone assisted oxo supplementation. J. Mater. Res. **11**, 1953–1959 (1996)
34. Goel, S.C., Chiang, M.Y., Gibbons, P.C., Buhro, W.E.: New chemistry for the sol-gel process: Acetone as a new condensation reagent. Mater. Res. Soc. Symp. Proc. **271**, 3–13 (1992)
35. Guerbet, M.: Action of alcohols on the sodium derivatives of other alcohols. Compt. Rend. **135**, 172–175 (1902)
36. Guerbet, M.: Sur trois alcools primaires nouveaux resultant de la condensation du benzylate de sodium avec les alcools propylique, butylique et isoamylique. Compt. Rend. **146**, 1405–1407 (1908)
37. Guo, G., Whitesell, J.K., Fox, M.A.: Synthesis of TiO_2 photocatalysts in supercritical CO_2 via non-hydrolytic route. J. Phys. Chem. B **109**, 18,781–18,785 (2005)
38. Haffad, D., Kameswari, U., Bettahar, M.M., Chambellan, A., Lavalley, J.C.: Reduction of benzaldehyde on metal oxides. J. Catal. **172**, 85–92 (1997)
39. Hong, Z.S., Cao, Y., Deng, J.F.: A convenient alcohothermal approach for low temperature synthesis of CuO nanoparticles. Mater. Lett. **52**, 34–38 (2002)
40. Hubert-Pfalzgraf, L.G., Daniele, S., Decams, J.M., Vaissermann, J.: The quest for single-source precursors for $BaTiO_3$ and $SrTiO_3$. J. Sol-Gel Sci. Technol. **8**, 49–53 (1997)

41. Hyeon, T., Lee, S.S., Park, J., Chung, Y., Na, H.B.: Synthesis of highly crystalline and monodisperse maghemite nanocrystallites without size-selection process. J. Am. Chem. Soc. **123**, 12,798–12,801 (2001)
42. Inoue, M., Kominami, H., Inui, T.: Novel synthetic method for the catalytic use of thermally stable zirconia: Thermal decomposition of zirconium alkoxides in organic media. Appl. Catal., A **97**, L25–L30 (1993)
43. Inoue, M., Kominami, H., Otsu, H., Inui, T.: Synthesis of microcrystalline titania in organic media. Nippon Kagaku Kaishi pp. 1364–1366 (1991)
44. Inoue, M., Otsu, H., Kominami, H., Inui, T.: Synthesis of yttrium-aluminum-garnet by the glycothermal method. J. Am. Ceram. Soc. **74**, 1452–1454 (1991)
45. Inoue, M., Tanino, H., Kondo, Y., Inui, T.: Formation of microcrystalline alpha-alumina by glycothermal treatment of gibbsite. J. Am. Ceram. Soc. **72**, 352–353 (1989)
46. Ivanda, M., Music, S., Popovic, S., Gotic, M.: XRD, Raman and FT-IR spectroscopic observations of nanosized TiO_2 synthesized by the sol-gel method based on an esterification reaction. J. Mol. Struct. **481**, 645–649 (1999)
47. Jansen, M., Guenther, E.: Oxide gels and ceramics prepared by a nonhydrolytic sol-gel process. Chem. Mater. **7**, 2110–2114 (1995)
48. Joo, J., Kwon, S.G., Yu, J.H., Hyeon, T.: Synthesis of ZnO nanocrystals with cone, hexagonal cone, and rod shapes via non-hydrolytic ester elimination sol-gel reactions. Adv. Mater. **17**, 1873–1877 (2005)
49. Joo, J., Kwon, S.G., Yu, T., Cho, M., Lee, J., Yoon, J., Hyeon, T.: Large-scale synthesis of TiO_2 nanorods via nonhydrolytic sol-gel ester elimination reaction and their application to photocatalytic inactivation of E. coli. J. Phys. Chem. B **109**, 15,297–15,302 (2005)
50. Joo, J., Yu, T., Kim, Y.W., Park, H.M., Wu, F., Zhang, J.Z., Hyeon, T.: Multigram scale synthesis and characterization of monodisperse tetragonal zirconia nanocrystals. J. Am. Chem. Soc. **125**, 6553–6557 (2003)
51. Jun, Y.W., Casula, M.F., Sim, J.H., Kim, S.Y., Cheon, J., Alivisatos, A.P.: Surfactant-assisted elimination of a high energy facet as a means of controlling the shapes of TiO_2 nanocrystals. J. Am. Chem. Soc. **125**, 15,981–15,985 (2003)
52. Kang, E., Park, J., Hwang, Y., Kang, M., Park, J.G., Hyeon, T.: Direct synthesis of highly crystalline and monodisperse manganese ferrite nanocrystals. J. Phys. Chem. B **108**, 13,932–13,935 (2004)
53. Karmaoui, M., Sá Ferreira, R.A., Mane, A.T., Carlos, L.D., Pinna, N.: Lanthanide-based lamellar nanohybrids: Synthesis, structural characterization, and optical properties. Chem. Mater. **18**, 4493–4499 (2006)
54. Kessler, V.G.: Molecular structure design and synthetic approaches to the heterometallic alkoxide complexes. Chem. Commun. pp. 1213–1222 (2003)
55. Kessler, V.G., Nikitin, K.V., Belokon, A.I.: A new argument in favor of the ether elimination mechanism: Formation of acetals on action of molybdenum alkoxides on carbonyl compounds. Polyhedron **17**, 2309–2311 (1998)
56. Kickelbick, G., Holzinger, D., Brick, C., Trimmel, G., Moons, E.: Hybrid inorganic-organic core-shell nanoparticles from surface- functionalized titanium, zirconium, and vanadium oxo clusters. Chem. Mater. **14**, 4382–4389 (2002)
57. Lafond, V., Mutin, P.H., Vioux, A.: Non-hydrolytic sol-gel routes based on alkyl halide elimination: toward better mixed oxide catalysts and new supports: Application to the preparation of a SiO_2-TiO_2 epoxidation catalyst. J. Mol. Catal. A: Chem. **182-183**, 81–88 (2002)
58. Lee, K., Seo, W.S., Park, J.T.: Synthesis and optical properties of colloidal tungsten oxide nanorods. J. Am. Chem. Soc. **125**, 3408–3409 (2003)
59. Li, X.L., Peng, Q., Yi, J.X., Wang, X., Li, Y.: Near monodisperse TiO_2 nanoparticles and nanorods. Chem. Eur. J. **12**, 2383–2391 (2006)
60. Li, Z., Chen, H., Bao, H.B., Gao, M.Y.: One-pot reaction to synthesize water-soluble magnetite nanocrystals. Chem. Mater. **16**, 1391–1393 (2004)

61. Li, Z., Sun, Q., Gao, M.Y.: Preparation of water-soluble magnetite nanocrystals from hydrated ferric salts in 2-pyrrolidone: Mechanism leading to Fe_3O_4. Angew. Chem. Int. Ed. **44**, 123–126 (2005)
62. Makowski, P., Rothe, R., A., T., Niederberger, M., Goettman, F.: Chlorine borrowing: An efficient method for an easier use of alcohols as alkylation agents. Green Chem. **11**, 34–37 (2009)
63. Mazdiyasni, K.S., Lynch, C.T., Smith, J.S.: Preparation of ultra-high-purity submicron refractory oxides. J. Am. Ceram. Soc. **48**, 372–375 (1965)
64. Meerwein, H., Schmidt, R.: Ein neues Verfahren zur Reduktion von Aldehyden und Ketonen. Liebigs Ann. Chem. **444**, 221–238 (1925)
65. Niederberger, M.: Nonaqueous sol-gel routes to metal oxide nanoparticles. Acc. Chem. Res. **40**, 793–800 (2007)
66. Niederberger, M., Bartl, M.H., Stucky, G.D.: Benzyl alcohol and titanium tetrachloride: A versatile reaction system for the nonaqueous and low-temperature preparation of crystalline and luminescent titania nanoparticles. Chem. Mater. **14**, 4364–4370 (2002)
67. Niederberger, M., Bartl, M.H., Stucky, G.D.: Benzyl alcohol and transition metal chlorides as a versatile reaction system for the nonaqueous and low-temperature synthesis of crystalline nano-objects with controlled dimensionality. J. Am. Chem. Soc. **124**, 13,642–13,643 (2002)
68. Niederberger, M., Garnweitner, G.: Nonaqueous synthesis of barium titanate nanocrystals in acetophenone as oxygen supplying agent. Mater. Res. Soc. Symp. Proc. **879E**, Z9.8.1–Z9.8.5 (2005)
69. Niederberger, M., Garnweitner, G.: Organic reaction pathways in the nonaqueous synthesis of metal oxide nanoparticles. Chem. Eur. J. **12**, 7282–7302 (2006)
70. Niederberger, M., Garnweitner, G., Ba, J., Polleux, J., Pinna, N.: Nonaqueous synthesis, assembly and formation mechanisms of metal oxide nanocrystals. Int. J. Nanotechnol. **4**, 263–281 (2007)
71. Niederberger, M., Garnweitner, G., Pinna, N., Antonietti, M.: Nonaqueous and halide-free route to crystalline $BaTiO_3$, $SrTiO_3$, and $(Ba,Sr)TiO_3$ nanoparticles via a mechanism involving C-C bond formation. J. Am. Chem. Soc. **126**, 9120–9126 (2004)
72. O' Lenick Jr., A.J.: Guerbet chemistry. J. Surfactants Deterg. **4**, 311–315 (2001)
73. Pande, K.C., Mehrotra, R.C.: Titanium salts of mono-carboxylic acids. 2. Reaction of titanium isopropoxide and ethoxide with acetic anhydride. Z. Anorg. Allg. Chem. **290**, 95–100 (1957)
74. Park, J., Kang, E., Son, S.U., Park, H.M., Lee, M.K., Kim, J., Kim, K.W., Noh, H.J., Park, J.H., Bae, C.J., Park, J.G., Hyeon, T.: Monodisperse nanoparticles of Ni and NiO: Synthesis, characterization, self-assembled superlattices, and catalytic applications in the suzuki coupling reaction. Adv. Mater. **17**, 429–434 (2005)
75. Pinna, N., Garnweitner, G., Antonietti, M., Niederberger, M.: Non-aqueous synthesis of high-purity metal oxide nanopowders using an ether elimination process. Adv. Mater. **16**, 2196–2200 (2004)
76. Pinna, N., Garnweitner, G., Antonietti, M., Niederberger, M.: A general nonaqueous route to binary metal oxide nanocrystals involving a C-C bond cleavage. J. Am. Chem. Soc. **127**, 5608–5612 (2005)
77. Pinna, N., Garnweitner, G., Beato, P., Niederberger, M., Antonietti, M.: Synthesis of yttria-based crystalline and lamellar nanostructures and their formation mechanism. Small **1**, 112–121 (2005)
78. Pinna, N., Grancharov, S., Beato, P., Bonville, P., Antonietti, M., Niederberger, M.: Magnetite nanocrystals: Nonaqueous synthesis, characterization, and solubility. Chem. Mater. **17**, 3044–3049 (2005)

79. Ponndorf, W.: Der reversible Austausch der Oxydationsstufen zwischen Aldehyden oder Ketonen einerseits und primären oder sekundären Alkoholen anderseits. Angew. Chem. **39**, 138–143 (1926)
80. Que, J., Ishimura, Y., Nagato, M.: Nippon Kagaku Kaishi p. 256 (1996)
81. Rauwel, E., Clavel, G., Willinger, M.G., Rauwel, P., Pinna, N.: Non-aqueous routes to metal oxide thin films by atomic layer deposition. Angew. Chem., Int. Ed. **47**, 3592–3595 (2008)
82. Rauwel, E., Willinger, M.G., Ducroquet, F., Rauwel, P., Matko, I., Kiselev, D., Pinna, N.: Carboxylic acids as oxygen sources for the atomic layer deposition of high-κ metal oxides. J. Phys. Chem. C **112**, 12,754–12,759 (2008)
83. Reetz M. T., P.R.: Erythro selective aldol condensation using titanium enolates. Tetrahedron Lett. **22**, 4691–4694 (1981)
84. Ritala, M., Kukli, K., Rahtu, A., Raisanen, P.I., Leskela, M., Sajavaara, T., Keinonen, J.: Atomic layer deposition of oxide thin films with metal alkoxides as oxygen sources. Science **288**, 319–321 (2000)
85. Rozes, L., Steunou, N., Fornasieri, G., Sanchez, C.: Titanium-oxo clusters, versatile nanobuilding blocks for the design of advanced hybrid materials. Monatsh. Chem. **137**, 501–528 (2006)
86. Sanchez, C., Soler-Illia, G., Ribot, F., Lalot, T., Mayer, C.R., Cabuil, V.: Designed hybrid organic-inorganic nanocomposites from functional nanobuilding blocks. Chem. Mater. **13**, 3061–3083 (2001)
87. Schubert, U.: Organically modified transition metal alkoxides: Chemical problems and structural issues on the way to materials syntheses. Acc. Chem. Res. **40**, 730–737 (2007)
88. Shiner, V.J., Whittaker, D.: The mechanism of the Meerwein-Ponndorf-Verley reaction. J. Am. Chem. Soc. **85**, 2337–2338 (1963)
89. Soler-Illia, G.J.d.A.A., Scolan, E., Louis, A., Albouy, P.A., Sanchez, C.: Design of meso-structured titanium oxo based hybrid organic-inorganic networks. New J. Chem. **25**, 156–165 (2001)
90. Spasova, M., Radetic, T., Sobal, N., Hilgendorff, M., Wiedwald, U., Farle, M., Giersig, M., Dahmen, U.: Structure and magnetism of Co and CoAg nanocrystals. Mater. Res. Soc. Symp. Proc. **721**, 195–200 (2002)
91. Sreekumar, R., Pillai, C.N.: Reactions of benzyl alcohol and dibenzyl ether over zeolites. Catal. Lett. **19**, 281–291 (1993)
92. Steunou, N., Bonhomme, C., Sanchez, C., Vaissermann, J., Hubert-Pfalzgraf, L.G.: A tetranuclear niobium oxo acetate complex. Inorg. Chem. **37**, 901–910 (1998)
93. Steunou, N., Ribot, F., Boubekeur, K., Maquet, J., Sanchez, C.: Ketones as an oxolation source for the synthesis of titanium-oxo-organo clusters. New J. Chem. **23**, 1079–1086 (1999)
94. Steunou, N., Robert, F., Boubekeur, K., Ribot, F., Sanchez, C.: Synthesis through an in situ esterification process and characterization of oxo isopropoxo titanium clusters. Inorg. Chim. Acta **279**, 144–151 (1998)
95. Tang, J., Redl, F., Zhu, Y., Siegrist, T., Brus, L.E., Steigerwald, M.L.: An organometallic synthesis of TiO_2 nanoparticles. Nano Lett. **5**, 543–548 (2005)
96. Tang, Z.Y., Ozturk, B., Wang, Y., Kotov, N.A.: Simple preparation strategy and one-dimensional energy transfer in cdte nanoparticle chains. J. Phys. Chem. B **108**, 6927–6931 (2004)
97. Trentler, T.J., Denler, T.E., Bertone, J.F., Agrawal, A., Colvin, V.L.: Synthesis of TiO_2 nanocrystals by nonhydrolytic solution-based reactions. J. Am. Chem. Soc. **121**, 1613–1614 (1999)
98. Turova, N.Y., Kessler, V.G., Kucheiko, S.I.: Molybdenum and tungsten (VI) bimetallic alkoxides: Decomposition accompanied by dialkylether elimination. Polyhedron **10**, 2617–2628 (1991)

99. Veibel, S., Nielsen, J.I.: On the mechanism of the Guerbet reaction. Tetrahedron **23**, 1723–1733 (1967)
100. Verley, A.: Bull. Soc. Chim. Fr. **37**, 537 (1925)
101. Vioux, A.: Nonhydrolytic sol-gel routes to oxides. Chem. Mater. **9**, 2292–2299 (1997)
102. Wang, C., Deng, Z., Li, Y.: The synthesis of nanocrystalline anatase and rutile titania in mixed organic media. Inorg. Chem. **40**, 5210–5214 (2001)
103. Wang, C., Deng, Z.X., Zhang, G.H., Fan, S.S., Li, Y.D.: Synthesis of nanocrystalline TiO_2 in alcohols. Powder Technol. **125**, 39–44 (2002)
104. Willinger, M., Neri, G., Rauwel, E., Bonavita, A., Micali, G., Pinna, N.: Vanadium oxide sensing layer grown on carbon nanotubes by a new atomic layer deposition process. Nano Lett. **8**, 4201–4204 (2008)
105. Woo, K., Hong, J., Ahn, J.P., Park, J.K., Kim, K.J.: Coordinatively induced length control and photoluminescence of $W_{18}O_{49}$ nanorods. Inorg. Chem. **44**, 7171–7174 (2005)
106. Woo, K., Hong, J., Choi, S., Lee, H.W., Ahn, J.P., Kim, C.S., Lee, S.W.: Easy synthesis and magnetic properties of iron oxide nanoparticles. Chem. Mater. **16**, 2814–2818 (2004)
107. Yin, M., Gu, Y., Kuskovsky, I.L., Andelman, T., Zhu, Y., Neumark, G.F., O'Brien, S.: Zinc oxide quantum rods. J. Am. Chem. Soc. **126**, 6206–6207 (2004)
108. Yin, M., Wu, C.K., Lou, Y., Burda, C., Koberstein, J.T., Zhu, Y., O'Brien, S.: Copper oxide nanocrystals. J. Am. Chem. Soc. **127**, 9506–9511 (2005)
109. Zhang, L., Djerdj, I., Cao, M., Antonietti, M., Niederberger, M.: Nonaqueous sol-gel synthesis of nanocrystalline $InNbO_4$ visible light photocatalyst. Adv. Mater. **19**, 2083–2086 (2007)
110. Zhang, Z., Liu, S., Chow, S., Han, M.Y.: Modulation of the morphology of ZnO nanostructures via aminolytic reaction: from nanorods to nanosquamas. J. Phys. Chem. B **22**, 6335–6340 (2006)
111. Zhang, Z., Zhong, X., Liu, S., Li, D., Han, M.: Aminolysis route to monodisperse titania nanorods with tunable aspect ratio. Angew. Chem. Int. Ed. **44**, 3466–3470 (2005)
112. Zhong, Z., Ang, T.P., Luo, J., Gan, H.C., Gedanken, A.: Synthesis of one-dimensional and porous TiO_2 nanostructures by controlled hydrolysis of titanium alkoxide via coupling with an esterification reaction. Chem. Mater. **17**, 6814–6818 (2005)

Chapter 6
Assembly

6.1 Introduction

Controlled assembly and positioning of nanoparticles in desired locations and across extended length scales is a key step in the design of integrated materials with advanced functions. The motivation to use nanoparticles as building blocks lies in the fact that size alone is not the defining parameter of many nanoscale-derived effects [28]. Also the interaction between the nanoparticles as a consequence of the spatial arrangement plays a role in determining the properties, opening-up unique and exciting opportunities for the creation of novel materials with unprecedented structures [70, 4]. Consequently, the future challenge of Nanoscience lies in the shift of focus from the size- and shape-controlled synthesis of nanoparticles towards the exploration of collective properties associated with their geometrical organization.

On the way to the development of a chemistry of organized matter [16], one can follow basically two strategies (although in some cases this division gets blurred): i) synthesis of nanoparticles and assembly into supraparticulate architectures in one step, and ii) synthesis and arrangement into 1-, 2- or 3-dimensional arrays in two consecutive steps.

Following the first strategy, Nature has given us many examples of how complex morphologies with hierarchical order can be achieved at ambient temperature and pressure and nearly neutral pH values through controlling crystallization and aggregation [16, 51]. Crystallization of biominerals does often not just proceed through ion-by-ion or single-molecule attachment to a nucleus, as proposed by classical crystallization theory, but also by a modular nanobuilding block route [14]. Interestingly, in spite of the nanoparticulate subunits and the finely carved appearance some of these biominerals are single crystals, which means that the primary particles are aligned and interconnected along a common crystallographic axis during their formation. These so-called mesocrystals are composed of individual nanoparticles, usually stabilized with organic ligands, which are aligned in a crystallographically oriented way, resulting not only in scattering properties that are similar to a single crystal, but often also in well facetted crystal habits. Fusion of high-energy crystal faces of the nanoparticulate constituents via oriented attachment could then lead to single crystals.

In this chapter we will exclusively discuss the very few examples of oriented attachment and mesocrystal formation for metal oxides in nonaqueous systems, which represent just a small part of this new, but rapidly growing research area. For an extensive overview of oriented attachment, mesocrystals and non-classical crystallization mechanisms the interested reader is referred to some recent reviews [14, 34, 76, 72, 29] and to an excellent textbook [15]. Of course, nanoparticles can also be assembled during synthesis without forming mesocrystals. Although these polycrystalline materials can have fascinating and complex morphologies, too, they will not be discussed in this chapter, due to the fact that their formation is in the most cases not based on a rational and predictive synthesis planning. By varying the reaction conditions, e.g., solvents and/or surfactants, it is possible to get different shapes and architectures, but without any possibility to predict the final morphology. As a matter of fact, rational planning of nanoparticle assemblies during synthesis seems to be even more challenging than that of nanoparticles with specific size and shape.

The second strategy, nanoparticle synthesis and assembly in two consecutive steps, will be discussed mainly based on superlattice formation. Most of the work on superlattices has been performed on metal and II-VI semiconductor nanocrystals, which are not part of this book. We will focus our discussion on just a few examples involving metal oxide nanoparticles. The assembly principles are, however, rather similar. Readers interested in getting a broad overview of nanoparticle assembly including all kinds of nanoparticles are referred to a large number of reviews [70, 4, 64, 17, 31, 49, 65, 53, 62, 36, 36, 10, 42, 68, 43] and textbooks [23, 48, 26]. The final part of this chapter is also dedicated to the second strategy and presents an example of how metal oxide nanoparticles can be arranged into fully crystalline mesoporous materials using an evaporation induced self-assembly process with block copolymers as templates.

6.2 Oriented Attachment and Mesocrystals

Mesocrystals are a special case of colloidal crystals that are produced via crystallization and aggregation basically in one step. However, the more frequently applied route to colloidal crystals involves a two step process, in which first monodisperse nanoparticles with desired size, shape and surface functionality are synthesized, followed by their assembly into superlattices using physical methods such as solvent evaporation [17, 31, 49], molecular cross-linking [53], programmed recognition [35, 18], or application of external magnetic [1, 38] or electric fields [59, 50]. Interestingly, in selected cases also these superlattices are built-up of oriented nanocrystals, which makes them another type of mesocrystals (cf. also Section 6.3) [76].

Crystal coarsening has classically been described as the growth of large particles at the expense of smaller ones (Ostwald ripening). The thermodynamic driving force for this process is the minimization of the surface energy. However, if the surface energy and the solubility of the material is low, other coarsening mechanisms like oriented attachment come into play. Oriented attachment describes the spontaneous self-organization of adjacent particles in such a way that they share a common crystallographic orientation, followed by fusion of these particles at a planar interface [39]. This process is particularly relevant in the nanocrystalline regime with relatively high specific surface areas, where bonding between the particles allows the system to win a substantial amount of energy by eliminating high-energy surfaces by crystallographic fusion as well as entropy by the release of surface-attached molecules [7, 3].

Oriented attachment processes in organic solvents were found for the formation of quasi one-, two- and three-dimensional structures. In contrast to aqueous media, in which many examples for the formation of 1D oxidic structures exist [40, 41, 25, 44, 47], in organic solvents only very few systems were reported. An almost classical example (and therefore we present it here, although strictly spoken it cannot be regarded as nonhydrolytic) is the formation of ZnO nanorods via oriented attachment of spherical nanocrystals that were obtained from zinc acetate in ethanol and KOH [37]. Increase of the particle concentration by evaporation of the solvent, followed by a heating step under reflux conditions induced the oriented attachment of the nanoparticles into pearl-chain-like structures, already with perfectly aligned lattice planes (Figure 6.1a), finally resulting in the formation of single crystalline nanorods (Figure 6.1b). The bottlenecks between adjacent particles were presumably filled up by the conventional mechanism of dissolution and growth of monomers.

Colloidal ceria nanocrystals, about 4 nm in diameter and synthesized from cerium nitrate using alkylamines as bases and Poly(vinylpyrrolidone) PVP as stabilizer, self-organize into four types of monolayer patterns [58]. Depending on the type of alkylamine and the concentration of PVP either isolated particles, short chain-like, pearl-necklace-like, or dendritic aggregates were found on the TEM grid. Figure 6.1c shows pearl-necklace-like aggregates. A closer look at this structures reveals that they are partly single-crystalline (Figure 6.1d). Interestingly, only PVP-containing colloidal ceria solutions formed such superstructures, which means that PVP favored the process of oriented aggregation. The authors explained this counterintuitive result with the bridge-linking role of PVP [58]. One PVP molecule is, due to its large size, able to link several particles together.

MnO multipods represent another example of quasi-1D oriented attachment [79]. $Mn(oleate)_2$ in trioctylamine and in the presence of oleic acid yielded almost exclusively hexapods with uniform size and shape (Figure 6.1e). Anisotropic assembly of MnO nanocrystals with the highly symmetric rocksalt structure is unexpected and was explained on the basis of induced dipole-dipole interaction. The formation of a rather strong dipole moment in a cubic crystal lattice was explained mainly by shape anisotropy [52]. A more detailed discussion of the MnO multipods and their characterization is provided in Section 7.2.3.

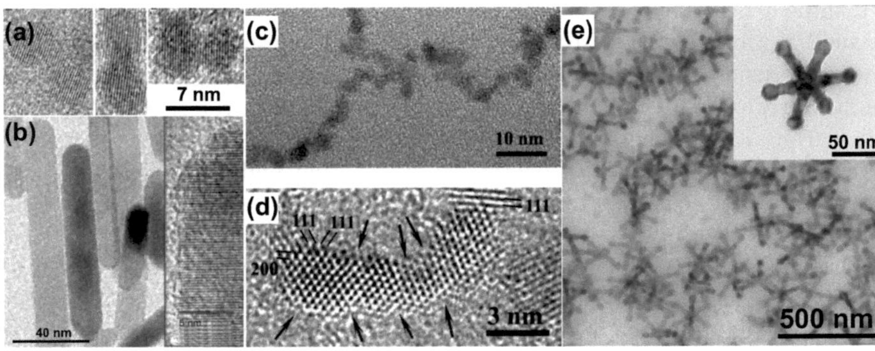

Fig. 6.1. a HRTEM images of ZnO nanoparticle assemblies, **b** TEM image of ZnO nanorods (right) together with a HRTEM image of a part of a nanorod (left). Images taken from Ref. [37] with permission of Wiley-VCH. **c** TEM image of pearl-necklace-like ceria nanoparticle chains, **d** HRTEM image of a single-crystalline part of such an assembly. Images taken from Ref. [58] with permission of the American Chemical Society. **e** TEM image of MnO multipods (inset: MnO hexapod)

In addition to oriented attachment processes in one dimension, also 2D assembly has been reported. Tungstite nanoplatelets that were synthesized by the "benzyl alcohol route" [45] are, according to electron diffraction measurements, perfectly single-crystalline. However, detailed HRTEM measurements of the individual nanoplatelets showed that they exhibited an internal composite structure, which means that every platelet is formed by fusion of a large number of small crystallites, just a few nanometers in size. The inner structure becomes visible as there is some misalignment of the orientation of these primary building blocks with respect to each other, leading to defects, which are additionally indicated by diffuse spots in the power spectrum. Details of the structural characterization are discussed in Section 7.2.3.

Oriented attachment can also occur in all three dimensions, leading to architectures that nicely fulfil the definition of mesocrystals. 5 nm-sized iron oxide nanocrystals synthesized in dimethylformamide in the presence of PVP aggregate to flower-like structures with diameters of 40–50 nm (Figure 6.2a) [74]. The TEM image of such a nanoflower clearly reveals the particulate nature (Figure 6.2b). Nevertheless, the corresponding fast Fourier transform pattern demonstrates its monocrystallinity (Figure 6.2c). Similar nanoflowers were reported for other metal oxides such as In_2O_3, CoO, MnO and ZnO based on a simple, but powerful strategy called limited ligand protection (LLP) [32, 33]. The idea was adapted from Kotov et al., who reported that CdTe nanocrystals formed 1D nanostructures through oriented attachment, when the excess of stabilizing ligand was removed [63]. By reducing the amount of stabilizing ligand to the domain of limited ligand protection, i.e., to the point at which the primary nanocrystals are insufficiently protected, three-dimensional, flower-like agglomerates were formed. Obviously, this approach can be applied to metal oxides with different compositions, crystal structures, and physical (magnetic and electronic) properties [33]. The synthesis involved the reaction between a metal carboxylate, consisting of a long-chain fatty acid, and 1-octadecyl alcohol in 1-octadecene. Figure 6.2d shows the temporal evolution of the morphology of the In_2O_3 nanocrystals from nanodots to nanoflowers [32]. In the first step, indium oxide nanodots formed as soon as the esterification reaction between the carboxylate species and the alcohol produced the water required for the hydrolysis reaction. With increasing reaction temperature, the nanodots grew further and at 250°C they agglomerate to the flower-like architectures. When the reaction temperature was increased to 270°C, the nanoflowers lost their pronounced shape and became more spherical. The single crystalline nature of the nanoflowers was confirmed by HRTEM investigations. Figure 6.2e shows such a micrograph with visible lattice fringes running across the entire flower structure, proving oriented attachment of several particles. Although gaps between the primary particles are clearly visible, they did not interfere with the lattice orientation. Accordingly, these nanoflowers correspond exactly to the definition of a mesocrystal, consisting of stabilized nanobuilding blocks, which are crystallographically oriented with respect to each other, resulting in single-crystalline scattering properties. The size of the nanoflowers can be adjusted in the range of 15 to 60 nm by varying the reaction conditions. It is interesting to note that in this case oriented attachment occurred randomly, i.e., no specific attachment axis was found, which means that the driving force must be more or less homogenous in all three dimensions. Therefore, the authors proposed that in addition to a potential dipole moment, also the minimization of the overall surface energy might represent an important driving force for oriented attachment [32]. The LLP approach seems to be a fascinating strategy for the design of complex 3D morphologies due to its applicability to a broad spectrum of colloidal nanocrystals independent of any electric or magnetic dipole moment [33]. However, the key issue is to identify the critical ligand protection point [32]. Above the critical ligand protection point, isolated and non-agglomerated nanodots will be formed, whereas below the critical point the nanodots become unstable and aggregate into 3D objects.

Several other examples of 3D architectures consisting of oriented nanoparticulate building blocks were reported, including $MnO/Co_{1-x}Mn_xO$ [71], CeO_2 [75], or ZnO [12]. Latest developments in the field of nonaqueous oxidic mesocrystal formation are the use of ionic liquids (although containing traces of water) as solvents [27], or microwave-irradiation as tool for tuning the size of the colloidal assemblies [20]. In the latter case, a zinc acetate solution in diethylene glycol (DEG) was heated in the microwave. The high polarizability of DEG makes this solvent an excellent microwave absorbing agent, thus leading to a high heating rate and significantly shorter reaction time. The obtained mesocrystals are spherical (Figure 6.2f) and single-crystalline according to electron diffraction (Figure 6.2f, inset). The HRTEM image in Figure 6.2g further supports the high crystallographic orientation of the nanoparticulate constituents. Most interestingly, the mesocrystals are quite monodisperse and their size can be tuned in the range of 57 to 274 nm [20].

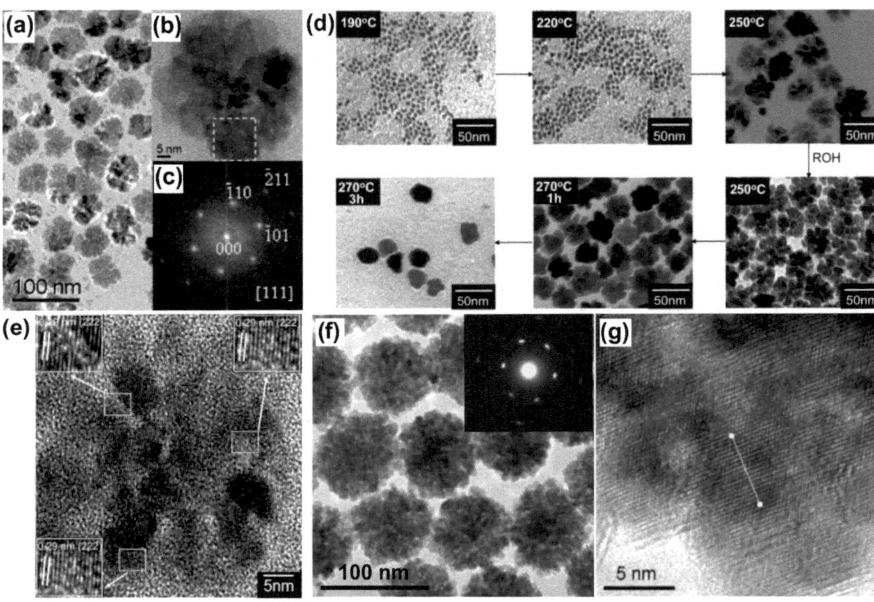

Fig. 6.2. **a** TEM overview image of hematite nanoflowers, **b** TEM image of one nanoflower and **c** FFT pattern of a part of the nanoflower. Images from Ref. [74] with permission of the American Chemical Society. **d** Temporal evolution of the morphology of In_2O_3 nanocrystals observed by TEM during the synthesis at 250°C, **e** HRTEM image of one nanoflower. Images reproduced from Ref. [32] with permission of the American Chemical Society. **f** TEM image of ZnO mesocrystals (inset: electron diffraction pattern of one particle) prepared by a microwave-polyol process, **g** HRTEM image of a part of one particle proving the orientation of the lattice fringes. Images taken from Ref. [20] with permission of Wiley-VCH

6.3 Superlattices

The use of nanoparticles as "artifical atoms" for the assembly of new materials provides the unique opportunity to combine the inherent functionality of the nanoparticles with potential collective properties arising from their interaction. Considering, how many different nanoparticles have been prepared in the last few years, and how many possibilities one has to arrange them, it is obvious that such a modular approach is highly attractive for the preparation of materials with programmable chemical and physical properties [61].

Usually, ordered arrangements of nanoparticles are classified by their dimensionality, including one-dimensional nanoparticle chains, two-dimensional arrays, and three-dimensional superlattices (also called superstructures or supercrystals). One-dimensional metal oxide nanoparticle chains are mainly reported in the context of oriented attachment processes (cf. Section 6.2), and therefore they are not further discussed here.

Two-dimensional ordering on surfaces can be achieved by spin or dip coating, and by the Langmuir-Blodgett (LB) technique [57, 22]. LB usually involves the deposition of the nanoparticles onto water. Recently, this technique has been extended to nonaqueous systems [2], which are generally more suitable, because many nanoparticles exhibit hydrophobic surfaces due to adsorbed surfactants. However, in the most cases, two-and three-dimensional superlattices are prepared by drop coating, i.e., by spreading a colloidal solution containing the nanoparticles on a solid substrate, usually a TEM grid, and subsequent evaporation of the solvent. The observation of superlattices can at least be traced back to the work of Bentzon et al., who reported the formation of hematite superlattices with hcp structure on the TEM grid [8, 9]. 20 years later, 2D assemblies of monodisperse metal oxide nanoparticles on TEM grids have become a rather common pattern (cf. for example Chapter 3.2 and 3.6), and therefore we will restrict our discussion to 3D superlattices.

We will present superlattices built up either by one or two kinds of nanoparticles that vary in size and/or composition. Furthermore, some of the processes are not just based on spontaneous self-assembly, but involve the use of a magnetic field to achieve long range ordering.

The simplest superlattices just consist of nanoparticles having all the same (spherical) size and composition. In this case, the packing rules are quite simple and face-centered cubic (fcc) and hexagonal close-packed (hcp) structures with maximum packing density of about 0.74 are formed [30, 38].

It is obvious that 3D packing is critically affected by the shape and anisotropy of the nanocrystals [21, 19]. Slight variation from the spherical morphology can result in different types of superlattices. Truncated octahedral magnetite nanocrystals, for example, assemble into a body-centered cubic (bcc) superlattice, and most interestingly the nanoparticles are crystallographically aligned within the superstructure [73]. The superlattices were prepared by placing a drop of Fe_3O_4 nanocrystals dispersed in hexane on a TEM copper grid and subsequent evaporation of the solvent. The obtained superlattices are composed of three kinds of pattern (Figure 6.3a-c). The corresponding fast Fourier transform (FFT) patterns in Figure 6.3d-f reveal the three projections along [001], [101], and [111] of the bcc structure. The electron diffraction patterns match with Fe_3O_4 with the intriguing feature that the diffraction spots clearly point to a single crystal (Figure 6.3g-i). Obviously, all the truncated octahedral nanoparticles in the superlattice have the same orientation (Figure 6.3j).

In the case of Zheng et al., the area covered by the well-ordered superstructure is typically in the range of 200 nm x 200 nm [73]. However, long-range order and macroscopic dimensions have been achieved by the application of an external magnetic field on a toluene-based dispersion of superparamagnetic oleate-capped maghemite nanocubes, resulting in defect-free superlattices with a very high degree of orientational order (Figure 6.3k) [1]. The structure of the particle arrays drastically altered from a mosaic-like pattern, when no magnetic field was used, to a superlattice with a very high degree of both translational and orientational order by subjecting the nanocube dispersion to a magnetic field of moderate strength (Figure 6.3l). Interestingly, the magnetic field has to be applied just for a very short period (less than 2 min) during the initial stage of the drying-mediated self-assembly process. The dimensions of the ordered superlattices can reach up to 10 μm. The individual nanocubes are separated from each other by a constant distance of 3.7 nm, which means that the oleate molecules on the surface are slightly interdigitated (Figure 6.3m). Comparison of the single crystal-like diffraction spots of the electron diffraction pattern on the atomic scale, corresponding to the [001] direction of maghemite (Figure 6.3n), with the mesoscale electron diffraction pattern (Figure 6.3o) proves that the crystallographic axes of the superlattices are identical to the orientation of the individual maghemite nanocubes. The jagged edge of the superlattice (Figure 6.3p), similar to a single crystal grown from solution, suggests that the superlattice grows by a slow crystallization process.

In addition to the preparation of superlattices over extended length scales and with a high degree of orientational order, superlattices with a well-defined size and shape represent another hardly explored challenge. The most promising results in this direction were achieved by using either oil-in-water microemulsions [6] or nanoparticle-micelles in ethylene glycol [77, 78]. Although two-phase systems shall not be discussed in this book, it is still worth mentioning that these approaches gave access to fascinating supercrystalline colloidal particles with good control over size and composition.

In comparison to superlattices just consisting of one kind of nanoparticulate building blocks, the assembly rules become much more complex, when two types of nanocrystals are mixed together. It is obvious that combination of several kinds of nanoparticles with distinctly different properties provides a unique opportunity for the bottom-up design of "metamaterials", i.e., materials with properties arising from the controlled interaction and mutual influence of their components [46].

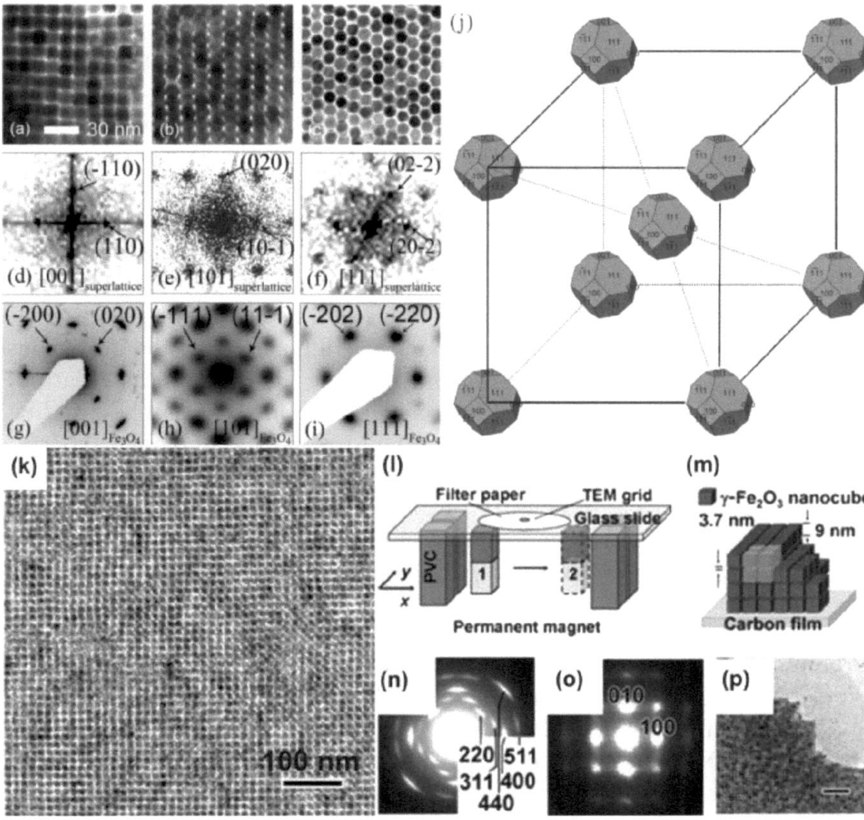

Fig. 6.3. **a**–**c** TEM image of the [001], [101], and [111] projections of the magnetite nanoparticle superlattices, respectively. **d**–**f** Corresponding FFT and **g**–**i** electron diffraction pattern of the superlattices shown in **a**–**c**. **j** A schematic of the bcc unit of the superlattice. Images taken from Ref. [73] with permission of Wiley-VCH. **k** TEM image of a magnetic-field-induced self-assembly of maghemite nanocubes into an oriented superlattice. **l** Schematic presentation of the experimental set-up. **m** Schematic presentation of the superlattice from nanocubes. **n**–**o** Atomic and mesoscale electron diffraction pattern, respectively, obtained from **k**. **p** TEM image obtained from the edge of the superlattice in **k** (Scale bar: 100 nm). Images taken from Ref. [1] with permission of The National Academy of Sciences of the USA

Purely oxidic binary superlattices have, to the best of our knowledge, not yet been reported and therefore we will discuss some examples, in which at least one nanoparticulate builing block comprises a metal oxide. Shortly after the publication of Zeng et al., reporting the combination of FePt and Fe_3O_4 nanocrystals for a 3D assembly [69], Redl et al. presented the self-assembly of PbSe semiconductor quantum dots and magnetic γ-Fe_2O_3 nanocrystals into a precisely ordered three-dimensional superlattice [46]. The idea to use specific size ratios to direct the assembly of the magnetic and semiconducting nanoparticles into AB_{13} or AB_2 superlattices represented the beginning of a fantastic journey through the great structural diversity of binary nanoparticle superlattices [55, 56].

Although only a few superlattice structures have been predicted to be thermodynamically stable, Shevchenko et al. demonstrated the formation of more than 15 different binary nanoparticle superlattices by combining semiconducting, metallic and magnetic nanocrystals [55, 56]. In many cases, several superlattice structures formed simultaneously on the same substrate under identical experimental conditions, i.e., the same nanoparticle mixture assembled into superlattices with different stoichiometry and packing symmetry. This observation underlines on the one hand the difficulty to get homogeneous superlattices over a large length scale, on the other hand shows the great potential of a modular self-assembly approach at the nanoscale. The building blocks, monodisperse PbS, PbSe, $CoPt_3$, Fe_2O_3, Au, Ag, and Pd nanocrystals stabilized with amphiphilic molecules with long alkyl chains, were dispersed either in toluene, toluene/tetrachloroethylene or toluene/chloroform. TEM grids or silicon nitride membranes as substrates were placed in a glass vial containing the colloidal solution. By evaporating the solvent under reduced pressure, binary nanoparticle superlattices with AB, AB_2, AB_3, AB_4, AB_5, AB_6, and AB_{13} stoichiometry with cubic, hexagonal, tetragonal, and orthorhombic symmetries have been identified, depending on the size and mixing ratio of the individual components [55]. Many of these arrays were found to be isostructural with NaCl, CuAu, AlB_2, $MgZn_2$, $MgNi_2$, Cu_3Au, Fe_4C, $CaCu_5$, CaB_6, $NaZn_{13}$, and cub-AB_{13} compounds, emphasizing the parallels between nanoparticle assembly and atomic scale crystal growth [56]. Figure 6.4 displays selected superlattices obtained from iron oxide nanoparticles as one, and Au (Figure 6.4a, k and l), and $CoPt_3$ (Figure 6.4g) nanoparticles as second constituents. The deposition of 5 nm gold and 13.4 nm γ-Fe_2O_3 nanocrystals yielded precisely ordered, large single domains of AB superlattices (Figure 6.4a), isostructural with NaCl (Figure 6.4b). The NaCl superlattices are usually represented by the (111) projection (Figure 6.4a and e), however smaller domains of the (100) projection were also observed (Figure 6.4a, upper inset, and c). Figures 6.4d and f depict the minimum number of layers in the (100) and (111) projections, respectively, necessary for the formation of the experimentally observed patterns. Figure 6.4g shows the (001) projection of an AlB_2-type superlattice assembled from 13.4 nm γ-Fe_2O_3 and 6.2 nm $CoPt_3$ nanocrystals, together with a 3D sketch of the AlB_2 unit cell (Figure 6.4h), its (001) plane (Figure 6.4i) and minimum number of layers in the (001) plane, leading to the observed patterns (Figure 6.4j). Changing the iron oxide to gold nanoparticle ratio to about 1:5, a superlattice with AB_6 stoichiometry (Figure 6.4k) isostructural with CaB_6 (Figure 6.4m) formed. View onto the (100) plane and the minimum number of layers in the (100) plane, leading to the experimentally observed patterns, are shown in Figure 6.4n and 6.4o, respectively.

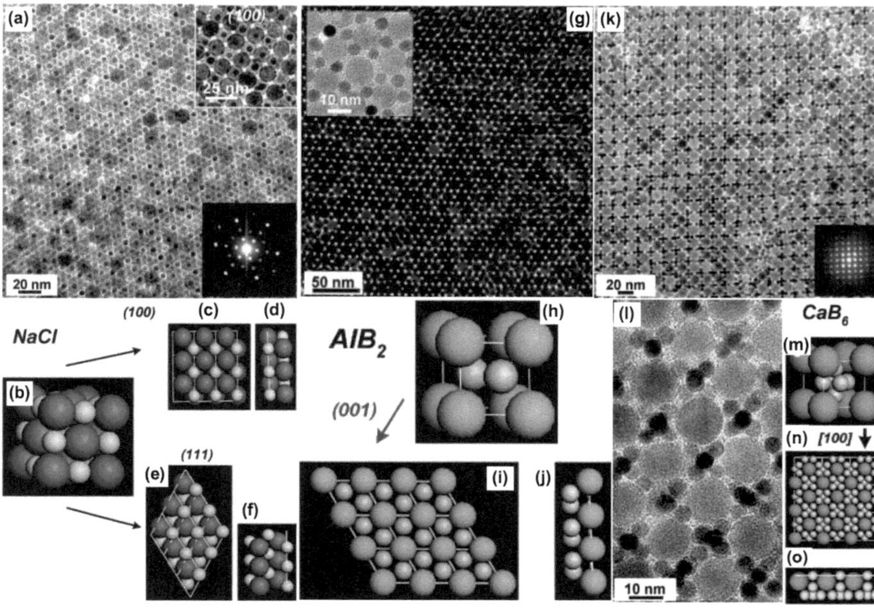

Fig. 6.4. a–o Binary superlattices using γ-Fe_2O_3 (13.4 nm) and Au (5 nm) or $CoPt_3$ (6.2 nm) nanocrystals as building blocks. **a** TEM micrograph of the AB superlattice of γ-Fe_2O_3 and Au nanocrystals, isostructural with NaCl, (111) projection. Upper inset: (100) projection. Lower inset: Electron diffraction pattern. **b** 3D sketch of the NaCl unit cell, **c**, **e** (100) and (111) planes, **d**, **f** minimum number of layers in the (100) and (111) projections necessary for the formation of the experimentally observed patterns. **g** TEM image of the (001) plane of γ-Fe_2O_3 and $CoPt_3$ nanocrystals, isostructural with AlB_2. Inset: Higher magnification. **h** 3D sketch of the AlB_2 unit cell, **i** (001) plane, **j** minimum number of layers in the (001) plane, leading to the observed pattern. **k** TEM micrograph of the AB_6 superlattice of γ-Fe_2O_3 and Au nanocrystals, isostructural with CaB_6, (100) projection. Inset: Small angle electron diffraction pattern from the (100) projection, **l** enlarged view. **m** 3D sketch of the CaB_6 unit cell, **n** (100) projection, **o** minimum number of layers in the (100) projection necessary for the formation of the experimentally observed pattern. Images reproduced from Ref. [56] with permission of the American Chemical Society

Many of these superlattices have packing densities smaller than 0.74, which indicates that the assembly properties of nanoparticles cannot simply be predicted on the basis of idealized hard spheres, thus ruling the packing density out as the major factor determining the stability of binary superlattices. Therefore, Shevchenko et al. proposed a model that demonstrates the importance of Coulomb, van der Waals, charge-dipole, and dipole-dipole interactions for the self-assembly of such complex nanoparticle superstructures [56]. Obviously, all the contribution of the different interactions is within the same order of magnitude. Without any doubts, the richness and diversity of binary superstructures still present a challenge to both the experimentalist and theorists to understand the principles that guide assembly at the nanoscale level [61, 56].

The ability to mix and match different nanocrystals and assemble them systematically into binary superlattices provides a unique route to metamaterials with precise uniformity of packing, stoichiometry, and rigorous control of the interparticle distance. In comparison to random mixtures of nanoparticles, the degree of ordering in 3D superlattices can have a pronounced effect on the collective optical, magnetic and electrical properties. Of course, binary superlattices do not represent the end of the story. As a matter of fact, ternary nanoparticle superlattices would even extend the range of accessible properties. A step in this direction was done by using core-shell nanoparticles as one of the building blocks. Assembly of single-phase gold with iron-iron oxide core-shell nanoparticles represents an elegant way to make quasi-ternary superlattices, at the same time avoiding the complexity of crystallizing three separate types of nanocrystals [54]. However, in spite of the success in the preparation of binary superlattices, studies on the collective properties are still scarce, and also their use as "real" materials in applications is still at the very beginning.

6.4 Mesoporous Materials

Since the discovery of mesoporous silica in 1992 [24], great efforts have been undertaken to prepare similar materials in the form of fully crystalline metal oxides [60, 67, 13]. Although partially crystalline mesoporous metal oxides have been reported ten years ago [66], the preparation of fully crystalline networks is still a major issue, mainly due to the fact that the as-synthesized amorphous mesostructures collapse during the annealing step required for inducing the crystallization process. One possibility to circumvent this problem is to use preformed and fully crystalline nanoparticles as nanobuilding blocks instead of molecular precursors. The experimental procedure for such mesoporous materials is closely related to the preparation of superlattices by evaporating the solvent of the colloidal solution, however with the difference that the colloidal solution contains a supramolecular template as porogen, e.g., surfactants or block copolymers (Figure 6.5a). One example, completely based on the use of organic solvents during the synthesis as well as during the assembly process, is the preparation of ordered mesoporous tin oxide with a fully crystalline framework [5]. In a first step, nearly monodisperse crystalline tin oxide nanoparticles with a diameter of 3.5 nm (Figure 6.5b, left panel) were prepared by reacting tin tetrachloride with benzyl alcohol (cf. also Chapter 4.1.3). The tin oxide nanoparticles could then directly be redispersed in tetrahydrofuran (THF), forming transparent and stable sols without any additional stabilizer and without any indication of particle agglomeration. Addition of a polybutadiene-(b)-poly(ethylene oxide) PB-PEO block-copolymer as template to the THF solution and subsequent evaporation of the solvent induced the cooperative assembly of the nanoparticles and the PB-PEO block-copolymer micelles (usually called evaporation-induced self-assembly [11]), finally resulting in the formation of large mesopores of 18–20 nm with an exceptionally high degree of ordering in terms of pore shape and 3D arrangement. The SEM and TEM images displayed in Figure 6.5b, right panel, prove the high quality of the mesostructure after thermal removal of the template at 500 °C, confirming that preformed nanocrystals are indeed able to preserve the mesostructure during calcination.

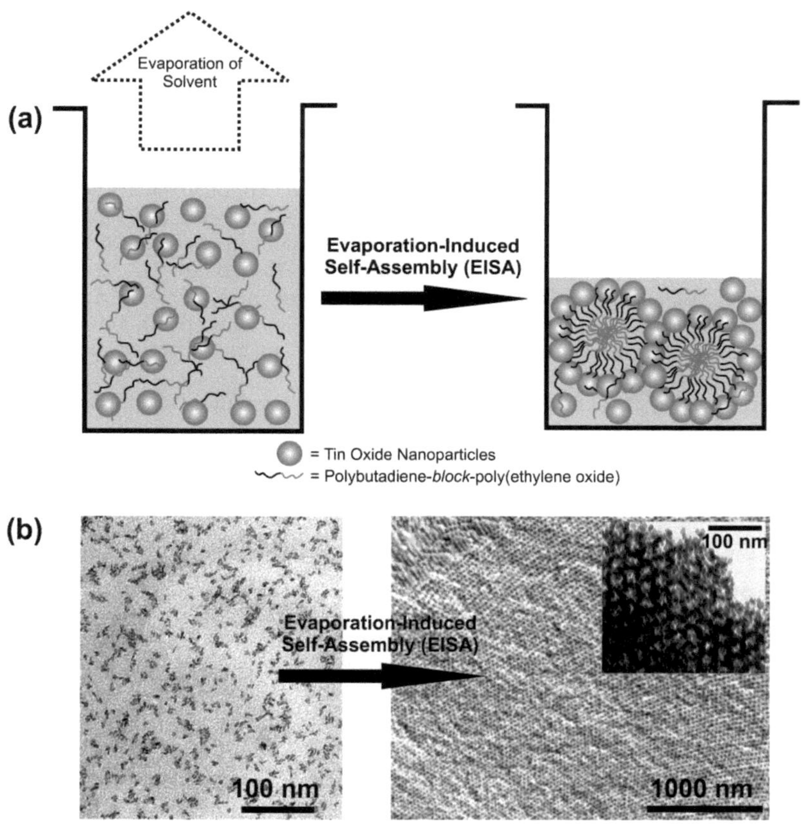

Fig. 6.5. **a** Schematic illustration of the evaporation-induced self-assembly of nanocrystals into mesoporous materials using block copolymers as templates. **b** TEM image of the starting colloidal solution of tin oxide nanocrystals (left) and SEM image of the final mesoporous material (right) after calcination (inset: TEM image of mesoporous tin oxide)

References

1. Ahniyaz, A., Sakamoto, Y., Bergstrom, L.: Magnetic field-induced assembly of oriented superlattices from maghemite nanocubes. Proc. Natl. Acad. Sci. U.S.A. **104**, 17,570–17,574 (2007)
2. Aleksandrovic, V., Greshnykh, D., Randjelovic, I., Fromsdorf, A., Kornowski, A., Roth, S.V., Klinke, C., Weller, H.: Preparation and electrical properties of cobalt-platinum nanoparticle monolayers deposited by the Langmuir-Blodgett technique. ACS Nano **2**, 1123–1130 (2008)
3. Alivisatos, A.P.: Biomineralization - Naturally aligned nanocrystals. Science **289**, 736–737 (2000)
4. Antonietti, M., Niederberger, M., Smarsly, B.: Self-assembly in inorganic and hybrid systems: beyond the moleclar scale. Dalton Trans. pp. 18–24 (2008)
5. Ba, J., Polleux, J., Antonietti, M., Niederberger, M.: Nonaqueous synthesis of tin oxide nanocrystals and their assembly into ordered porous mesostructures. Adv. Mater. **17**, 2509–2512 (2005)
6. Bai, F., Wang, D.S., Huo, Z.Y., Chen, W., Liu, L.P., Liang, X., Chen, C., Wang, X., Peng, Q., Li, Y.D.: A versatile bottom-up assembly approach to colloidal spheres from nanocrystals. Angew. Chem. Int. Ed. **46**, 6650–6653 (2007)
7. Banfield, J.F., Welch, S.A., Zhang, H.Z., Ebert, T.T., Penn, R.L.: Aggregation-based crystal growth and microstructure development in natural iron oxyhydroxide biomineralization products. Science **289**, 751–754 (2000)
8. Bentzon, M.D., Vanwonterghem, J., Morup, S., Tholen, A.: Ordered aggregates of ultrafine iron-oxide particles - super crystals. Philos. Mag. B **60**, 169–178 (1989)
9. Bentzon, M.D., Vonwonterghem, J., Tholen, A.: A magnetic super lattice. Ultramicroscopy **24**, 67–68 (1988)
10. Böker, A., He, J., Emrick, T., Russell, T.P.: Self-assembly of nanoparticles at interfaces. Soft Matter **3**, 1231–1248 (2007)
11. Brinker, C.J., Lu, Y.F., Sellinger, A., Fan, H.Y.: Evaporation-induced self-assembly: Nanostructures made easy. Adv. Mater. **11**, 579–585 (1999)
12. Buha, J., Djerdj, I., Niederberger, M.: Nonaqueous synthesis of nanocrystalline indium oxide and zinc oxide in the oxygen-free solvent acetonitrile. Cryst. Growth Des. **7**, 113–116 (2007)
13. Carreon, M.A., Guliants, V.V.: Ordered meso- and macroporous binary and mixed metal oxides. Eur. J. Inorg. Chem. pp. 27–43 (2005)
14. Cölfen, H., Antonietti, M.: Mesocrystals: inorganic superstructures made by highly parallel crystallization and controlled alignment. Angew. Chem. Int. Ed. **44**, 5576–5591 (2005)
15. Cölfen, H., Antonietti, M.: Mesocrystals and nonclassical crystallization. John Wiley & Sons Ltd., Chichester, England (2008)
16. Cölfen, H., Mann, S.: Higher-order organization by mesoscale self-assembly and transformation of hybrid nanostructures. Angew. Chem. Int. Ed. **42**, 2350–2365 (2003)
17. Collier, C.P., Vossmeyer, T., Heath, J.R.: Nanocrystal superlattices. Annu. Rev. Phys. Chem. **49**, 371–404 (1998)
18. Gianneschi, N.C., Masar, M.S., Mirkin, C.A.: Development of a coordination chemistry-based approach for functional supramolecular structures. Acc. Chem. Res. **38**, 825–837 (2005)
19. Glotzer, S.C., Solomon, M.J.: Anisotropy of building blocks and their assembly into complex structures. Nature Mater. **6**, 557–562 (2007)
20. Hu, X., Gong, J., Zhang, L., Yu, J.C.: Continuous size tuning of monodisperse ZnO colloidal nanocrystal clusters by a microwave-polyol process and their application for humidity sensing. Adv. Mater. **20**, 4845–4850 (2008)

21. Jana, N.R.: Shape effect in nanoparticle self-assembly. Angew. Chem. Int. Ed. **43**, 1536–1540 (2004)
22. Kinge, S., Crego-Calama, M., Reinhoudt, D.N.: Self-assembling nanoparticles at surfaces and interfaces. Chem. Phys. Chem. **9**, 20–42 (2008)
23. Kotov, N.A.: Nanoparticle assemblies and superstructures. CRC Press: Boca Raton (2006)
24. Kresge, C.T., Leonowicz, M.E., Roth, W.J., Vartuli, J.C., Beck, J.S.: Ordered mesoporous molecular-sieves synthesized by a liquid-crystal template mechanism. Nature **359**, 710–712 (1992)
25. Lee, E.J.H., Ribeiro, C., Longo, E., Leite, E.R.: Oriented attachment: an effective mechanism in the formation of anisotropic nanocrystals. J. Phys. Chem. B **109**, 20,842–20,846 (2005)
26. Lee, Y.S.: Self-assembly and nanotechnology: A force balance approach. John Wiley & Sons, Inc.: Hoboken, New Jersey (2008)
27. Li, Z.H., Gessner, A., Richters, J.P., Kalden, J., Voss, T., Kubel, C., Taubert, A.: Hollow zinc oxide mesocrystals from an ionic liquid precursor (ILP). Adv. Mater. **20**, 1279–1285 (2008)
28. Mann, S.: Life is a nanoscale phenomenon. Angew. Chem. Int. Ed. **47**, 5306–5320 (2008)
29. Meldrum, F.C., Cölfen, H.: Controlling mineral morphologies and structures in biological and synthetic systems. Chem. Rev. **108**, 4332–4432 (2008)
30. Murray, C.B., Kagan, C.R., Bawendi, M.G.: Synthesis and characterization of monodisperse nanocrystals and close-packed nanocrystal assemblies. Annu. Rev. Mater. Sci. **30**, 545–610 (2000)
31. Murray, C.B., Norris, D.J., Bawendi, M.G.: Synthesis and characterization of nearly monodisperse CdE (E=S, Se, Te) semiconductor nanocrystallites. J. Am. Chem. Soc. **115**, 8706–8715 (1993)
32. Narayanaswamy, A., Xu, H., Pradhan, N., Kim, M., Peng, X.: Formation of nearly monodisperse In_2O_3 nanodots and oriented-attached nanoflowers: Hydrolysis and alcoholysis vs pyrolysis. J. Am. Chem. Soc. **128**, 10,310–10,319 (2006)
33. Narayanaswamy, A., Xu, H., Pradhan, N., Peng, X.: Crystalline nanoflowers with different chemical compositions and physical properties grown by limited ligand protection. Angew. Chem. Int. Ed. **45**, 5361–5364 (2006)
34. Niederberger, M., Cölfen, H.: Oriented attachment and mesocrystals: Non-classical crystallization mechanisms based on nanoparticle assembly. Phys. Chem. Chem. Phys. **8**, 3271–3287 (2006)
35. Niemeyer, C.M., Simon, U.: DNA-based assembly of metal nanoparticles. Eur. J. Inorg. Chem. pp. 3641–3655 (2005)
36. Ozkan, C.S., Wang, Z.L.: Assembly at the nanoscale: Toward functional nanostructured materials. Small **2**, 1322–1323 (2006)
37. Pacholski, C., Kornowski, A., Weller, H.: Self-assembly of ZnO: From nanodots, to nanorods. Angew. Chem. Int. Ed. **41**, 1188–1191 (2002)
38. Park, J., Joo, J., Kwon, S.G., Jang, Y., Hyeon, T.: Synthesis of monodisperse spherical nanocrystals. Angew. Chem. Int. Ed. **46**, 4630–4660 (2007)
39. Penn, R.L., Banfield, J.F.: Imperfect oriented attachment: dislocation generation in defect-free nanocrystals. Science **281**, 969–971 (1998)
40. Penn, R.L., Banfield, J.F.: Oriented attachment and growth, twinning, polytypism, and formation of metastable phases: Insights from nanocrystalline TiO_2. Am. Mineral. **83**, 1077–1082 (1998)
41. Penn, R.L., Banfield, J.F.: Morphology development and crystal growth in nanocrystalline aggregates under hydrothermal conditions: Insights from titania. Geochim. Cosmochim. Acta **63**, 1549–1557 (1999)
42. Pileni, M.P.: Self-assembly of inorganic nanocrystals: Fabrication and collective intrinsic properties. Acc. Chem. Res. **40**, 685–693 (2007)

43. Pileni, M.P.: Supracrystals of inorganic nanocrystals: An open challenge for new physical properties. Acc. Chem. Res. **41**, 1799–1809 (2009)
44. Polleux, J., Pinna, N., Antonietti, M., Hess, C., Wild, U., Schlogl, R., Niederberger, M.: Ligand functionality as a versatile tool to control the assembly behavior of preformed titania nanocrystals. Chem. Eur. J. **11**, 3541–3551 (2005)
45. Polleux, J., Pinna, N., Antonietti, M., Niederberger, M.: Growth and assembly of crystalline tungsten oxide nanostructures assisted by bioligation. J. Am. Chem. Soc. **127**, 15,595–15,601 (2005)
46. Redl, F.X., Cho, K.S., Murray, C.B., O'Brien, S.: Three-dimensional binary superlattices of magnetic nanocrystals and semiconductor quantum dots. Nature **423**, 968–971 (2003)
47. Ribeiro, C., Lee, E.J.H., Longo, E., Leite, E.R.: Oriented attachment mechanism in anisotropic nanocrystals: A "polymerization" approach. Chem. Phys. Chem. **7**, 664–670 (2006)
48. Rogach, A.: Semiconductor nanocrystal quantum dots: Synthesis, assembly, spectroscopy and applications. Springer Verlag: Wien (2008)
49. Rogach, A.L., Talapin, D.V., Shevchenko, E.V., Kornowski, A., Haase, M., Weller, H.: Organization of matter on different size scales: Monodisperse nanocrystals and their superstructures. Adv. Funct. Mater. **12**, 653–664 (2002)
50. Ryan, K.M., Mastroianni, A., Stancil, K.A., Liu, H.T., Alivisatos, A.P.: Electric-field-assisted assembly of perpendicularly oriented nanorod superlattices. Nano Lett. **6**, 1479–1482 (2006)
51. Sanchez, C., Arribart, H., Giraud Guille, M.M.: Biomimetism and bioinspiration as tools for the design of innovative materials and systems. Nature Mater. **4**, 277–288 (2005)
52. Shanbhag, S., Kotov, N.A.: On the origin of a permanent dipole moment in nanocrystals with a cubic crystal lattice: Effects of truncation, stabilizers, and medium for CdS tetrahedral homologues. J. Phys. Chem. B **110**, 12,211–12,217 (2006)
53. Shenhar, R., Rotello, V.M.: Nanoparticles: Scaffolds and building blocks. Acc. Chem. Res. **36**, 549–561 (2003)
54. Shevchenko, E.V., Kortright, J.B., Talapin, D.V., Aloni, S., Alivisatos, A.P.: Quasi-ternary nanoparticle superlattices through nanoparticle design. Adv. Mater. **19**, 4183–4188 (2007)
55. Shevchenko, E.V., Talapin, D.V., Kotov, N.A., O'Brien, S., Murray, C.B.: Structural diversity in binary nanoparticle superlattices. Nature **439**, 55–59 (2006)
56. Shevchenko, E.V., Talapin, D.V., Murray, C.B., O'Brien, S.: Structural characterization of self-assembled multifuntional binary nanoparticle superlattices. J. Am. Chem. Soc. **128**, 3620–3637 (2006)
57. Shipway, A.N., Katz, E., Willner, I.: Nanoparticle arrays on surfaces for electronic, optical, and sensor applications. Chem. Phys. Chem. **1**, 18–52 (2000)
58. Si, R., Zhang, Y.W., You, L.P., Yan, C.H.: Self-organized monolayer of nanosized ceria colloids stabilized by poly(vinylpyrrolidone). J. Phys. Chem. B **110**, 5994–6000 (2006)
59. Smith, P.A., Nordquist, C.D., Jackson, T.N., Mayer, T.S., Martin, B.R., Mbindyo, J., Mallouk, T.E.: Electric-field assisted assembly and alignment of metallic nanowires. Appl. Phys. Lett. **77**, 1399–1401 (2000)
60. Soler-Illia, G.J.d.A.A., Sanchez, C., Lebeau, B., Patarin, J.: Chemical strategies to design textured materials: from microporous and mesoporous oxides to nanonetworks and hierarchical structures. Chem. Rev. **102**, 4093–4138 (2002)
61. Talapin, D.V.: LEGO materials. ACS Nano **2**, 1097–1100 (2008)
62. Tang, Z., Kotov, N.A.: One-dimensional assemblies of nanoparticles: Preparation, properties, and promise. Adv. Mater. **17**, 951–962 (2005)
63. Tang, Z.Y., Kotov, N.A., Giersig, M.: Spontaneous organization of single CdTe nanoparticles into luminescent nanowires. Science **297**, 237–240 (2002)

64. Wang, Z.L.: Structural analysis of self-assembling nanocrystal superlattices. Adv. Mater. **10**, 13–30 (1998)
65. Weller, H.: Synthesis and self-assembly of colloidal nanoparticles. Phyl. Trans. R. Soc. Lond. A **361**, 229–240 (2003)
66. Yang, P.D., Zhao, D.Y., Margolese, D.I., Chmelka, B.F., Stucky, G.D.: Generalized syntheses of large-pore mesoporous metal oxides with semicrystalline frameworks. Nature **396**, 152–155 (1998)
67. Yu, C.Z., Tian, B.Z., Zhao, D.Y.: Recent advances in the synthesis of non-siliceous mesoporous materials. Curr. Opin. Solid State Mater. Sci. **7**, 191–197 (2003)
68. Zabet-Khosousi, A., Dhirani, A.A.: Charge transport in nanoparticle assemblies. Chem. Rev. **108**, 4072–4124 (2008)
69. Zeng, H., Li, J., Liu, J.P., Wang, Z.L., Sun, S.H.: Exchange-coupled nanocomposite magnets by nanoparticle self-assembly. Nature **420**, 395–398 (2002)
70. Zhang, H., Edwards, E.W., Wang, D., Möhwald, H.: Directing the self-assembly of nanocrystals beyond colloidal crystallization. Phys. Chem. Chem. Phys. **8**, 3288–3299 (2006)
71. Zhang, H.T., Chen, X.H.: Morphology-controllable synthesis and characterization of hierarchical 3D $Co_{1-x}Mn_xO$ nanostructures. J. Phys. Chem. B **110**, 9442–9447 (2006)
72. Zhang, Q., Liu, S.J., Yu, S.H.: Recent advances in oriented attachment growth and synthesis of functional materials: Concept, evidence, mechanism, and its future. J. Mater. Chem. **19**, 191–207 (2009)
73. Zheng, R.K., Gu, H.W., Xu, B., Fung, K.K., Zhang, X.X., Ringer, S.P.: Self-assembly and self-orientation of truncated octahedral magnetite nanocrystals. Adv. Mater. **18**, 2418–2421 (2006)
74. Zheng, Y., Cheng, Y., Wang, Y., Bao, F., Zhou, L., Wei, X., Zhang, Y., Zheng, Q.: Quasicubic α-Fe_2O_3 nanoparticles with excellent catalytic performance. J. Phys. Chem. B **110**, 3093–3097 (2006)
75. Zhou, H.P., Zhang, Y.W., Mai, H.X., Sun, X., Liu, Q., Song, W.G., Yan, C.H.: Spontaneous organization of uniform CeO_2 nanoflowers by 3D oriented attachment in hot surfactant solutions monitored with an in situ electrical conductance technique. Chem. Eur. J. **14**, 3380–3390 (2008)
76. Zhou, L., O'Brien, P.: Mesocrystals: a new class of solid materials. Small **4**, 1566–1574 (2008)
77. Zhuang, J.Q., Wu, H.M., Yang, Y.A., Cao, Y.C.: Supercrystalline colloidal particles from artificial atoms. J. Am. Chem. Soc. **129**, 14,166–14,167 (2007)
78. Zhuang, J.Q., Wu, H.M., Yang, Y.G., Cao, Y.C.: Controlling colloidal superparticle growth through solvophobic interactions. Angew. Chem. Int. Ed. **47**, 2208–2212 (2008)
79. Zitoun, D., Pinna, N., Frolet, N., Belin, C.: Single crystal manganese oxide multipods by oriented attachment. J. Am. Chem. Soc. **127**, 15,034–15,035 (2005)

Chapter 7
Characterization

7.1 Introduction

In order to relate the physical properties to the size, shape and crystallinity of nanoobjects an accurate and detailed characterization has to be performed. The determination of the size and shape distribution of nanometer size particles can be addressed with several techniques like for example: analytical ultracentrifugation (AUC) [3], light scattering techniques, scanning electron microscopy (SEM) and transmission electron microscopy (TEM) [50, 51, 52, 19, 21] etc. The most common tools for the structural characterization of nanoobjects are high resolution TEM (HRTEM) and diffraction techniques such as electron diffraction (ED) and powder X-ray diffraction (XRD). HRTEM permits to directly visualize the atomic columns of a single particle and to determine its structure and possible structural defects. However, this method is not statistically applicable to a large amount of particles. Powder XRD measurements are able to overcome this limitation and provide a global information about the crystallinity of a sample, thus making it a perfect complementary technique to HRTEM. This chapter mainly focuses on the characterization of inorganic nanoobjects by electron microscopy and diffraction techniques. Additional useful and widely used characterization techniques such as Fourier transform infrared (FT-IR) and solid state nuclear magnetic resonance (SSNMR) will also be discussed in a particular example. In the first section (7.2) the transmission electron microscope and the basic operation modes are introduced together with some examples of structure determination by ED and HRTEM. The examples will be taken from nanocrystals synthesized in organic solvents. Afterward, the chapter discusses powder X-ray diffraction techniques on nanopowders and especially how to extract typical information related to the nanometric size and the crystallinity of nanoparticles (7.3). The advanced use of electron microscopy and diffraction techniques will be illustrated along with the structural determination of hollandite-type vanadium oxyhydroxide nanorods (7.4.1). Finally, the study of rare earth based organic-inorganic hybrid materials will be detailed as it requires the use of several techniques, in addition to electron microscopy and diffraction techniques, in order to obtain structural information (7.4.2).

7.2 Transmission Electron Microscopy

In general, the attainable resolution of an optical system is limited by the wavelength of the probing waves according to Abbes relation. For visible light the obtainable resolution is about 300 nm. The principle of a transmission electron microscope (TEM) is analogous to the one of an optical transmission microscope, but using an electron beam instead of visible light. However, due to the different nature of electrons and photons, the optical lenses have to be replaced by magnetic and electrostatic lenses. The big advantage of electrons is that their wavelength can be decreased simply by accelerating them across a high voltage. In 1932 Max Knoll and Ernst Ruska already proposed that when using electrons instead of light for imaging, one could achieve a resolution in the order of a few Å [43]. Indeed, resolution in this range was obtained about 40 years later. Nowadays, instruments reach resolutions down to the sub-Å region [1, 27]. A TEM provides highly coherent electrons of sufficient energy and allows to investigate a sample in terms of its morphology, structure and orientation at a very high spatial resolution. Imaging over a big range of magnifications offers the possibility to determine the size distribution of particles by measuring a few hundreds of them at low resolution and to determine the shape and structure of individual particles at higher resolution. In addition to imaging, a TEM also provides ideal conditions for analyzing the interactions taking place between the electron beam and the sample material. These interactions carry valuable information that can be retrieved by spectroscopic methods. Spatially resolved compositional analysis by X-ray emission spectroscopy (energy dispersive X-ray spectroscopy, EDXS) and local electronic structure measurements by electron energy loss spectroscopy (EELS) and energy filtered imaging at the nanometer scale are some of the techniques that are commonly applied in analytical TEM. In the following paragraph, the set-up and the basic operation modes of the instrument will briefly be introduced together with some examples chosen from nanocrystals synthesized in organic solvents.

7.2.1 The Operation Modes

Depending on the operation mode of the instrument, different contrast mechanisms are involved in the formation of the final image.

7.2.1.1 Low Magnification

In low magnification mode, contrast observed on the viewing screen is caused by electrons that have been lost from the imaged region either by large-angle scattering outside the objective aperture (contrast aperture) or as a result of energy loss and wavelength change in the specimen (mass-thickness contrast). Electrons that have lost some of their kinetic energy get focused on a plane far distant from the viewing screen. This out-of-focus "inelastic" or energy loss image only contributes a uniform background to the in-focus elastic image. In the so-called bright-field mode, the contrast aperture selects the direct, unscattered spot and excludes all electrons that were Bragg-scattered to higher angles, whereas in the dark-field mode the image is formed only from scattered electrons. The possibility to select electrons according to a certain range of scattering angle allows to point out sample areas with different scattering characteristics like crystal defects in mono-crystals or micro-crystals of different orientation in polycrystalline samples.

7.2.1.2 High Resolution

High resolution transmission electron microscopy (HRTEM) is one of the most powerful tools used for the characterization of nanomaterials. The most important contrast mechanism for high resolution electron microscopy is phase-contrast. If the specimen is suitably thin for HRTEM, i.e. less than about 50 nm, the interaction of the electron beam with the inner potential of the specimen predominantly causes phase shifts of parts of the electron wave front. The image interpretation in HRTEM is not straight forward. This is because in addition to the interaction of the electron beam with the specimen, also the effect of the imaging process by imperfect electron lenses has to be considered. The way how the microscope transfers waves with different relative phase shifts down through the column is defined by the so called contrast transfer function. It depends on the performance of the objective lens (spherical aberration), the acceleration voltage and the focus conditions. Further, chromatic aberrations, focal and energy spread of the beam and instabilities in the high tension and objective lens current play a role in the contrast formation. Assistance in the interpretation of HRTEM images is provided by image processing and image simulation. Some examples will be given later.

Even though HRTEM is the most used technique to determine the structure of a single isolated nanoparticle, the structure determination of small particles is often difficult and not always possible especially when the nanoparticles are very small (i.e. less than 2 nm in diameter), present complicated structures or crystallographic defects.

The structure determination of one isolated nanoparticle is in general performed in few steps. First of all the power spectrum (PS) of the high resolution image has to be calculated. The PS is the square of the Fourier transform of the 2D image. From the PS the netplane distances and plane angles can be obtained more easily than from the real images. After that the PS has to be compared to single crystal diffraction calculated for the possible structure for different orientations. This approach works well, if the particles studied do not present any defects and are crystallized in a known structure. If this is not the case a more detailed study is needed such as Fourier analysis and HRTEM simulations. However, this would go well beyond the aim of this chapter and only few examples will be given later.

In order to calculate the PS of a HRTEM image a square selection of the image of a power of 2 should be chosen (generally 256, 512 or 1024 pixels depending of the nanoparticle size and resolution of the image). Normally a resolution between 0.03 and 0.02 nm/pixel should be used to not lose structural information (Figure 7.1a). Furthermore, in order to avoid edge effects which could introduce reflections or lines which do not come from the crystal structure, a spherical mask with a Gaussian cut of the intensity should be applied (Figure 7.1b). Finally the PS of this masked image is calculated. It shows several well defined sharp spots (Fig 7.1c). The netplane distances are measured by using the following simple formula:

$$d_{hkl} = \frac{N_{pix-image} * P_{size}}{N_{pix}} \tag{7.1}$$

where $N_{pix-image}$ is the size of the quadratic frame of the PS, P_{size} is the pixel size in the real image, and N_{pix} is the number of pixels between the origin and the reflection hkl in the reciprocal space. The PS always show pairs of reflections hkl and $\overline{h}\overline{k}\overline{l}$ which are lying diametrical opposite around the origin.

Figure 7.1 shows only an explanatory example, more advanced studies may include: i) Masks of different sizes can be used in order to determine the structure at different positions. This is often useful when structural defects are present or when the particle to be studied is not a single crystal. ii) Fourier analysis which consist of selecting only interesting pair of reflections and subsequently apply a back Fourier transform. This technique produces a filtered image presenting less noise and information on the lattice planes chosen. This is useful for the study of non-perfect crystals, etc. Few examples of how to apply these techniques will be given below.

The use of a large amount of surfactants during synthesis produces metal oxide particle precipitates together with a non-negligible amount of organic material difficult to completely separate from the inorganic one. Usually, these organic surfactants prevent to acquire good quality HRTEM images. In fact, the surfactants burn under the electron beam reducing the stability of the particles during the measurements. This is generally not the case for nanoparticles produced in surfactant-free systems which present a very low content of organic compounds [25] and permit the acquisition of high quality HRTEM images like the ones presented in figure 7.2. For imaging, the particles were

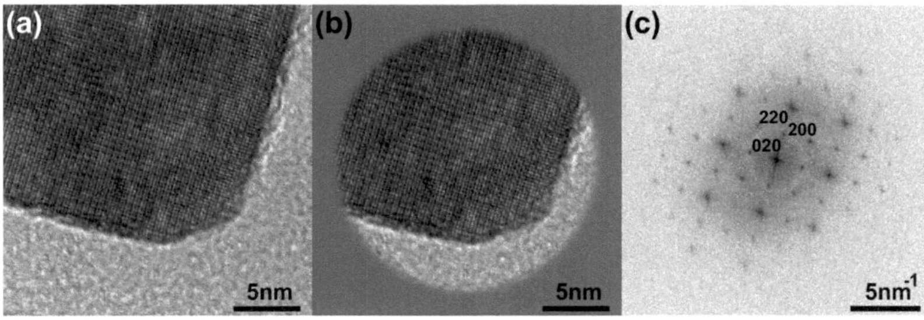

Fig. 7.1. **a** HRTEM image of 1024x1024 pixels from a part of an In_2O_3 cube-shaped nanoparticle, **b** masked image, **c** PS of the image in **b**

transferred to an amorphous carbon coated copper grid by depositing one drop of a dispersion of the as synthesized particles in ethanol onto it. Fig 7.2a depicts a part of a 20 nm In_2O_3 cubic nanocrystal [36], showing well defined edges and monocristalline behavior as demonstrated by the PS in insert. Such high crystallinity characteristic of nanoparticles produced by the "benzyl alcohol route" exists also for very small particles such as $Sn_{0.95}In_{0.05}O_x$ (Figure 7.2b) [24, 23]). Despite its extremely small size (~2 nm) this cubic particle shows a perfect crystalline order. Larger spherical particles like in the case of magnetite [35] (Figure 7.2c) are also perfect monocrystals having crystalline order even at their surface which presents sharp and well defined facets. The last panel shows the [010] zone axis of a Nb_2O_5 orthorhombic nanoplatelet [32] once more the crystalline order is remarkable especially for such an orthorhombic structure which is known to include many defects also in the bulk phase.

7.2.1.3 Electron Diffraction

In this mode, the electron microscope is used as a diffractometer. This is simply done by changing the excitation of the intermediate lens system such that instead of focusing on the first image plane of the objective lens, it is focused on the back focal plane of the objective which contains the Fraunhofer diffraction pattern. The projector lens then projects a magnified image of the diffraction pattern on the fluorescent screen. The easy switch over from imaging to diffraction and vice versa and the variability of the electron probe size is very helpful for analytical work. Small single crystals down to the size of 5 nm can be selected in image mode and their diffraction characteristic can be studied simply by changing to the diffraction mode. The arrangement and intensity of diffraction spots is related to the symmetry and dimension of the unit cell and depends on the orientation of the particle with respect to the incident beam. In order to resolve the lattice parameters and crystal symmetry of a particle, diffraction patterns have to be recorded from different

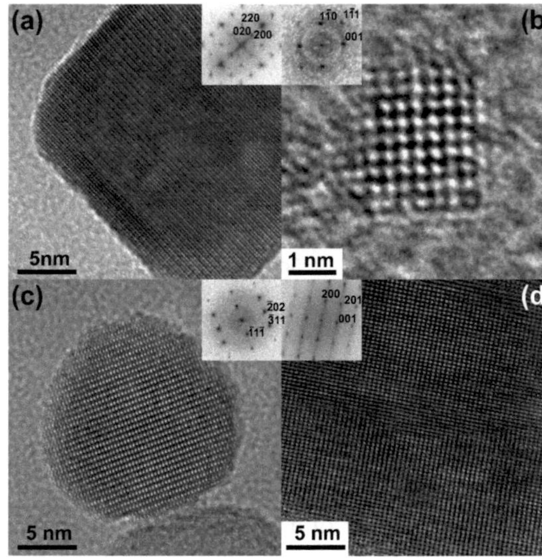

Fig. 7.2 **a** HRTEM image of a part of an In_2O_3 cube-shaped nanoparticle, **b** a 2 nm nanoparticle $Sn_{0.95}In_{0.05}O_x$, **c** a 16 nm nanoparticle of Fe_3O_4, **d** a part of a Nb_2O_5 nanoplatelet. Inserts show respective PS

crystallographic projections. This is usually done by tilting the sample to different observation directions with respect to the incident electron beam or in the case of a homogeneous sample, by selecting particles of different orientations. From the distance between diffracted spots, the plane spacings can be calculated by use of Bragg's law.

In Figure 7.3 an isolated nanoplatelet of tungstite [40] is presented together with its selected area ED (SAED). The TEM image of an isolated platelet shows a quadratic and well faceted shape (Figure 7.3a). The SAED pattern (Figure 7.3b) of this particle shows spots characteristic of tungstite ($WO_3{\cdot}H_2O$) oriented along the [001] direction, demonstrating that the particle is monocrystalline. However, the HRTEM study of these platelets shows (cf. Chapter 7.2.3) that they are far from being perfect monocrystals and that they are in fact built up by a large number of small crystallites of a few nanometers in size which are oriented almost but not exactly, parallel to one another.

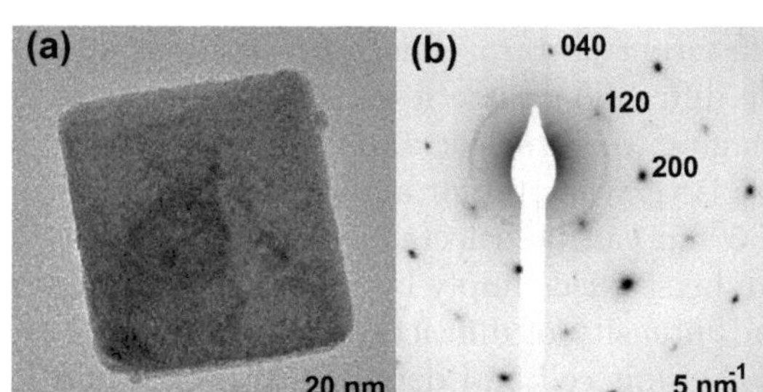

Fig. 7.3 **a** TEM overview image of a tungstite nanoplatelet, **b** SAED of the same platelet oriented along the [001] direction

7.2.1.4 Scanning Transmission Electron Microscopy

In the scanning mode (STEM), the specimen is scanned across the region of interest by using a very small probe of typically only a few Å in diameter. If the specimen region is sufficiently thin, most electrons are transmitted with relatively little lateral spreading, and a variety of signals from elastic and inelastic scattering events can be recorded, allowing high spatial resolution analysis and imaging. At each point, various types of signals discriminated by scattering angle and energy loss yield different structural and chemical information that can be captured simultaneously in different channels. Most commonly, energy dispersive X-ray spectroscopy (EDX), electron energy loss spectrometry (EELS) and high angle annular dark field detectors (HAADF) are applied. For a detailed introduction on STEM we refer to the book of R. J. Keyse, A. J. Garratt-Reed, P. J. Goodhew and G. W. Lorimer [18].

7.2.2 HRTEM Image Simulations

If the comparison between the extracted crystallographic data from SAED, HRTEM, PS does not allow an unambiguous determination of the structure of the nanocrystal, a more detailed study is required. In this case, the most used technique is the simulation of HRTEM images on the basis of the structure of model particles. The calculations are often performed with the multislice technique [47, 4, 48]. This approach includes several steps:

1. The construction of one or more atomic models of the nanocrystals
2. The simulation of the HRTEM image of these models
3. The calculation of the PS of the calculated HRTEM image
4. The comparison of the HRTEM images and the PS calculated with the data obtained from the experimental HRTEM

It should be noticed that the contrast of a HRTEM image strongly depends on the defocus, so particular care has to be taken in acquiring the HRTEM images.

In the following we will apply this technique to MnO nanoparticles synthesized in benzyl alcohol presenting an interesting superstructure which can be described as an ordered Mn vacancy cubic superstructure with the general formula of $Mn_{0.875}O_x$ and a lattice parameter of 8.888 Å [7].

The HRTEM image of a manganese oxide nanoparticle (Figure 7.4a) together with its PS (Figure 7.4b) show a peculiar feature associated with the spot B in the PS. The corresponding d-value equals 5.078 Å, which is exactly the double of spot A. This reflection can be indexed only with rational Miller indices 0.5 0.5 0.5. Such a configuration of spots in the PS is characteristic of the $[0\bar{1}1]$ zone axis orientation of the MnO nanocrystal displayed. The indexing of the PS with rational numbers points to a superstructure.

A careful look at the Fourier-filtered HRTEM image in Figure 7.4c clearly shows arrays of spots. One bright line of spots (denoted by arrowheads) alternates with a less bright one. The distance between two lines of spots with the same brightness is exactly the double (5.078 Å) of the distance between arrays of brightless bright spots. Such a regular arrangement of double arrays of spots found in the HRTEM image demonstrates the presence of a regular superstructure. The assigned rational Miller indices can be turned into

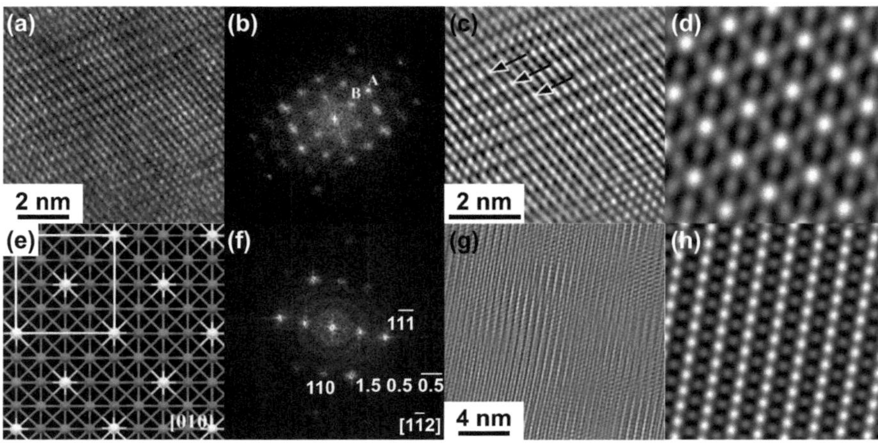

Fig. 7.4. **a** HRTEM image of a part of a single-crystalline MnO nanoparticle, **b** corresponding PS, **c** Fourier-filtered HRTEM image of part of **a** (arrows denote the lattice fringes corresponding to the superstructure), **d** simulated image based on the model shown in **e**, **f** PS of a [$1\bar{1}2$] HRTEM image of a single-crystalline MnO nanoparticle, **g** Fourier filtered HRTEM image from **f**, **h** simulated image based on the model shown in **e**

integers by simple multiplication by a factor of 2, implying that new integer Miller indices describing the structure of MnO with a doubled value of the unit cell parameter can be used. One possible superstructure constitutes the regular distribution of manganese or oxygen vacancies or interstitials in the MnO crystal matrix. It is unlikely that the oxygen vacancies are responsible for the superstructural ordering, because HRTEM is less sensitive to light atoms. Therefore, ordered Mn vacancies should be responsible for the superstructure. In order to prove this hypothesis HRTEM image simulation using the multislice techniques were performed. The superstructural model used in the simulations is displayed in Figure 7.4e in the projection along the b-axis. The manganese atoms are removed from the corners and centers of the faces in a simple doubled unit cell of MnO (yellow color), leaving behind an ordered system of Mn vacancies. A simple calculation shows that the Mn deficiency in such a model is 12.5%, i.e., the formula of the defective MnO structure is $Mn_{0.875}O_x$. On the basis of the proposed vacancy ordered superstructural model, the HRTEM image simulation was performed using the multislice method, and considering the experimental conditions of HRTEM imaging and specific zone axis. Taking the experimental parameters of microscope into account, the image is simulated at Scherzer focus (−68 nm) and at the sample thickness of 2.5 nm (Figure 7.4d). The bright contrast in the simulated image corresponds to the Mn vacancy column, because in HRTEM imaging the atom deficiency columns leave the signature as a bright contrast at this particular defocus. The visual inspection of the simulated image shows that it nicely matches with the Fourier-filtered HRTEM image of a MnO nanocrystal oriented in the $[0\bar{1}1]$ zone (Figure 7.4c). In this way, the HRTEM measurement corroborates the validity of the proposed superstructure model based on Mn vacancies ordering. To prove that this example is not an exceptional case the same approach was applied for the $[1\bar{1}2]$ orientation of the MnO crystal with respect to the electron beam. The corresponding PS (Figure 7.4f) also shows reflections that can be indexed only in terms of rational numbers, pointing again to the formation of a superstructure. The Fourier-filtered image (Figure 7.4g) clearly exhibits the interchange of two lattice fringes in the $[\bar{1}11]$ direction that are mutually shifted in the [110] direction by d_{220} (1.57 Å). The simulated image (Figure 7.4h) unequivocally confirms the superstructural model as it also fits nicely with the experimental one. It is important to highlight that the MnO superstructure is not detected by powder XRD but only with local probing (HRTEM and SAED) at the level of the single nanocrystal [7].

7.2.3 TEM Studies of Nanostructures by Oriented Attachment

Above we showed how to study highly crystalline nanocrystals synthesized in organic solvents, in the next selected examples we will show the structural properties of nanoparticles formed by the oriented attachment growth mechanism (cf. Chapter 6). This non-classical crystallization mechanism permits to produce anisotropic nanostructures. However, it often leads to the inclusion of defects where the primary building blocks epitaxially aggregate. Especially, we will show the particular features of crystals grown under this particular mechanism and how by using HRTEM techniques it is possible to obtain information on the growth mechanism. Furthermore, these systems are interesting for a microscopist because they often present interesting defects where the attachment of the primary building blocks occurs.

In Figure 7.5 an example of the study of anatase particles synthesized by the benzyl alcohol route is presented. In this case the anatase particles are functionalized with 2-amino-2-(hydroxymethyl)-1,3-propanediol $(HOCH_2)_3CNH_2$ (Trizma) during the synthesis and subsequently refluxed in water in order to arrange them into pearl-necklace structures [39, 38]. Figure

7.5a shows an imperfect oriented attachment along the [001] axis characterized by a small misorientation at the interface (black arrow). This is confirmed by the Power Spectrum (PS), which was calculated for the entire nanorod and shows doubled spots (Figure 7.5b). This effect is due to the fact that the lattice planes on both sides of the nanorod are misoriented leading to slightly tilted PS for the two parts of the rod. Further prove is given in Figure 7.5c, where the PS has been calculated only for the left part of the nanorod (white circle). In this case, the spots are no longer doubled. In order to get further information on the misorientation of the lattice plane at the attachment region Fourier analysis is performed. This technique consists on selecting a particular crystal direction on the PS by masking it. In the present case the PS of the entire image was masked in order to get information only from the (020) reflections. Subsequently, the inverse Fourier transform of the masked PS in calculated (Figure 7.5d). In this figure only the information concerning the lattice planes of the (020) family are present and thus, the interpretation of the image is drastically facilitated. For example, the presence of dislocation is clearly seen (black circles) and the misorientation is easily estimated to about 4° from the green and red lines drawn on the image (Figure 7.5d).

Another example of oriented attachment of primary particles leading to anisotropic nanostructures is given by the non-aqueous synthesis of tungstite nanoplatelets by the benzyl alcohol route [40]. The particles exhibit a platelet like shape and have sides ranging from 50 to 100 nm and thicknesses from 5 to 10 nm. Contrarily to the example of the TiO_2 nanorods, in this case the primary particles spontaneously assemble to form anisotropic structures without any specific molecule adsorbed at their surface and a post-synthetic treatment. SAED measured from isolated tungstite nanoplatelets prove their structure and were characteristic of monocrystals (Figure 7.3). However, by HRTEM (Figure 7.5e) the particle seems to be formed by a large number of small perfect crystallites of few nanometers in size which are oriented almost but not exactly, parallel to one another. The platelet shows a much wider but lower diffraction profile as compared to perfect crystals. This behavior is further confirmed by the PS of this HRTEM image (Figure 7.5f) showing reflection spots characteristic of the tungstite structure for a particle oriented along the [001] direction. These reflections are diffuse and reflect the fact that the little crystallites are not perfectly aligned with respect to each other. Unfortunately from a PS as well as from an electron diffraction pattern it is not possible to extract the primary crystallite sizes. In order to better characterize the growth mechanism of the platelets once more Fourier analysis is needed. The PS is masked in order to only keep the information about the (120) lattice planes (black circle in figure 7.5f). Thus, the inverse Fourier transform will only show information regarding those lattice planes (Figure 7.5g). Non-perfectly aligned lattice planes almost oriented in the same direction but not perfectly as in a single crystal runs along the whole selected zone. In addition many defects like dislocations could be observed as well as large zones where the attachment of the planes is slightly misoriented. These findings suggest that the platelets are indeed formed by oriented attachment of primary particles having an average size of around 5 nm which by directed aggregation form the final crystals with large number of defects.

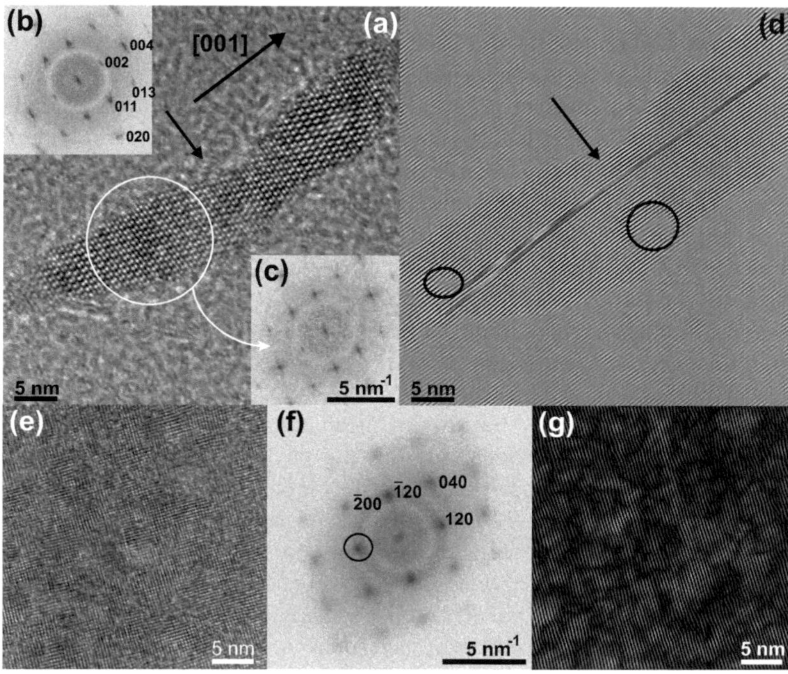

Fig. 7.5. **a** HRTEM of a part of a TiO_2 nanorod grown by oriented attachment, **b** PS of the entire image, **c** PS of a selected area marked with a white circle in **a**, **d** Fourier filtered image from the (020) reflections. **e** HRTEM of a part of a tungstite platelet, and **f** its PS. **g** Fourier analysis of the HRTEM image in **e**

MnO multipods were synthesized via a non-aqueous sol-gel route using $Mn(oleate)_2$ in n-trioctylamine [56]. In the presence of oleic acid, the synthesis yields almost exclusively hexapods homogeneous in size and shape. TEM study showed homogeneous NCs around 200 nm in size (Figure 7.6a). Most of the multipods appeared to be hexapods with an Oh symmetry even if some of the NCs displayed only 2, 3, 4 or 5 pods (Figure 7.6b-g). All of the NCs were single-crystalline as evidenced by electron diffraction and high resolution studies. Core and pods were found to be homogeneous in size and morphology, each arm being terminated by an arrow. However, they were not exempted from defects as demonstrated by electron diffraction and high resolution studies. SAED measured from isolated multipods (Figure 7.6h) showed patterns similar to the one presented in Figure 7.6i. The SAED showed sharp single spots characteristic of a single crystal of MnO (manganosite) oriented along the [001] direction with its arms parallel to the [100] and [010] directions respectively. Furthermore, it should be noted that the presence of 110 forbidden reflections points to the presence of a high density of stacking faults or defects. These findings were further proved by HRTEM. The image of the an arm-end (Figure 7.6j) showed well defined lattice planes. A magnification of the white squared zone and its PS (Figure 7.6k and l) confirmed the good crystallinity of the multipods and their growth orientation. Once again in the PS the presence of the formally forbidden 110 reflections point to the fact that the multipods are not perfect crystals and present a high density of defects. In fact, stacking faults and defects may also give rise to symmetric forbidden reflections because, due to a break of symmetry, the extinction rules of a particular reflection are not any more satisfied. Furthermore, the amount of defects will determine the intensity of the forbidden reflections and the diffuse character of the spots. This is what is occurring in the MnO multipods for which the amount of defects is high. This behavior is frequent for little particles and nanostructures, like the present ones, made by oriented attachment in which the amount of defects introduced by the imperfect attachment is often very high.

In order to further prove the growth mechanism the authors studied the kinetics of the reaction by characterizing the particles after only 30 min of growth. In this case they observed that the particles adopt a truncated octahedral shape of about 40-50 nm which is in good agreement with chain end diameter and some particles coalesced to form aggregates as shown in Figure 7.6m. The structure was also manganosite as confirmed by SAED (Figure 7.6n). Two independent works published only few months later reported similar multipods or dumbbells synthesized by an analogous route [55, 28]. The authors of one of them [55] proposed that both sides of the arms of the dumbbells are not flat but show pearl-chain-like structure. This suggests that they are formed by aggregation of diamond-shaped particles, thus supporting the hypothesis of oriented attachment growth mechanism by aggregation of MnO diamond-like nanoparticles. Such diamond-shaped particles were also obtained at slightly different reaction conditions supporting the hypothesis. The structural properties of these dumbbells were also similar to the multipods reported above. In fact, by HRTEM they observe that the nanocrystals are highly monocrystalline and confirm that the dumbbells exhibit cubic rock salt structure and that the arms are parallel to the [100] and [010] directions. Furthermore, the PS of a part of a crystal shows again the formally forbidden 110 reflections due to the large amount of defects [55].

The few examples given above and the growing number of articles published on the subject suggests that detailed structural studies and especially HRTEM studies are needed for the precise characterization of the oriented attachment growth mechanism.

In conclusion, from the cases studied above the reader gets familiar with some of the possible approaches necessary to fully characterize the structures formed by this particular type of particle-based crystal growth.

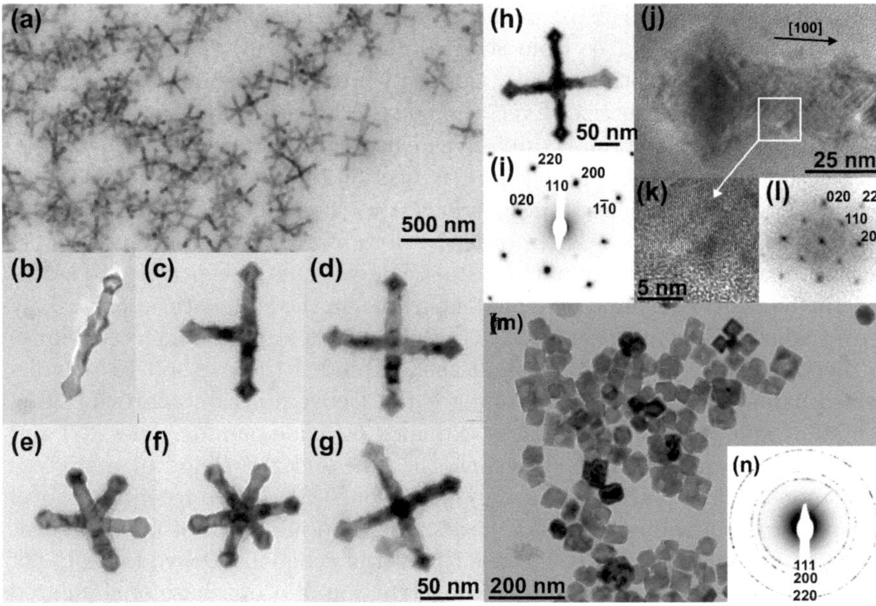

Fig. 7.6. **a** TEM overview image of the MnO multipods, **b** a bipod, **c** a tripod, **d** a tetrapod, **e** a pentapod, **f** a hexapod, **g** a branched pentapod, **h** a typical tetrapod, **i** SAED showing also the forbidden 110 reflections, **j** HRTEM of a single arm, **k** magnification of the white square, **l** its PS indicating once more the presence of the forbidden 110 reflections, **m** TEM overview image of the MnO nanocrystals grown for 30 min only and **n** its related SAED

7.3 Powder X-ray Diffraction

As previously demonstrated, TEM studies permit to directly visualize the atomic columns of a single particle and to determine its structure, its defects and so on. However, this method is not statistically applicable to a large amount of particles. Powder XRD measurements are able to overcome such a limitation and give a global information about the crystallinity of the whole sample, thus make it a perfect complementary technique to HRTEM. The Scherrer equation is commonly used to determine the average crystal size from the powder XRD measurement [29]:

$$B(2\theta) = \frac{K\,\lambda}{L\,cos\theta} \tag{7.2}$$

Where $B(2\theta)$ is the full width in radians at half maximum intensity of the powder pattern peak and L is the cube edge dimension and K a constant which depends on the particle morphology and it varies between 0.9 and 1 [29]. In the original work Scherrer evaluated the constant to $K = 2\,[(ln\ 2)/\pi]^{1/2} = 0.94$ for a cubic particle [45]. Such an equation is only valid under the following approximations: i) The shape of the crystal is well defined (cubic, spherical, etc.), ii) the crystal is free from strains, defects, etc. iii) the system should be monodisperse. In other words, the peak broadening should only be due to the small crystallite size. Even though the Scherrer equation should be applied for isotropic crystals only, it can be used for anisotropic materials, with acceptable results, if for each hkl-reflection the L value is interpreted as an average crystal dimension perpendicular to the reflecting plane.

More exact methods exist in order to determine particle size and shape, crystallinity, defects, etc. in nanocrystals. For example, it is possible to calculate the intensity scattered by a model particle, in the kinematic approach, by the so called Debye function [6, 53, 11]. A few groups applied such techniques to metallic clusters [49, 14, 54] and then extended the method to polydispersed nanocrystalline systems calling it the Debye function analysis (DFA) [10, 12]. Later on, this method was extended to oxide nanoparticles [37] and applied to characterize the structure, the size and the shape of many oxide nanoparticles synthesized in organic solvents [31]. More recently a novel approach was introduced for size and shape determination of nanoparticles based on calculations of XRD powder patterns using the Debye formula [2]. The methodology was based on the construction of a database of simulated XRD powder patterns calculated from all possible morphologies (sizes and shapes) of the considered nanoparticle system. Comparison between the calculated XRD diagram database and experimental XRD powder patterns is then performed by means of calculation of a Rietveld agreement factor.

The Debye formula describes the intensity distribution, spherically averaged over the reciprocal space:

$$I_N(b) = \sum_{n,m\neq n}^{N} f_n f_m \frac{sin(2\pi b r_{nm})}{2\pi b r_{nm}} \tag{7.3}$$

Where $b = \frac{1}{d} = \frac{2sin\theta}{\lambda}$, λ is the wavelength and 2θ the scattering angle. The sum runs over all the pair distances r_{nm} of the atoms labeled by n,m with scattering amplitudes f_n, f_m in a cluster of N atoms. The formula includes the following characteristics which are significant advantages compared to the Scherrer equation: i) It is a general equation valid for any form of matter in which there is a random orientation: gases, liquids, amorphous solids, and crystalline powders. ii) There is no limitation on the number of different kinds of atoms in the sample. The main disadvantage stays on the number of terms, which have to be calculated, that increases proportional to the sixth order of the linear dimension of the particle, thus, restricting the applicability to particles smaller than 5–6 nm. By applying efficient algorithms larger particles might be calculated too [13].

$BaTiO_3$ nanocrystals synthesized in benzyl alcohol [26] (Figure 7.7) can be taken as a model example for demonstrating how the Debye equation permits to precisely elucidate the crystal structure of nanocrystals. All diffraction peaks in Figure 7.7 can be assigned to the $BaTiO_3$ phase without any indication of other crystalline by-products such as barium carbonate or titanium dioxide. The cubic and tetragonal perovskite structures of $BaTiO_3$ are very similar in terms of lattice distances and thus, difficult to distinguish at the nanoscale. The powder pattern in the 2θ = 40-50° region is characteristic for the presence of either the cubic or tetragonal $BaTiO_3$ structure, and the splitting of the 200 into tetragonal 200 and 002 reflections at about 45° is observed. In the present case, the reflections are too broad to discriminate between the two crystal modifications due to the small particle size. This can be easily confirmed by calculations of the Debye equation for the two possible structures for spherical, monodisperse particles with diameters of 6 nm. There are only minor differences between the cubic (dotted line) and the tetragonal (dashed line) calculated patterns. Moreover, the pattern of the tetragonal crystal modification does not display any splitting of reflections due to the small size. Therefore, in this case XRD cannot discriminate between the two very similar structures. Nevertheless, calculations of the Debye equations confirm the very good crystallinity of the perovskite particles, their spherical shape and the absence of systematic structural defects. In fact, the calculated position, the broadening and the relative intensity of the experimental and calculated diffractions fit very well. Furthermore, the average size extracted from Debye equation studies corresponds to the one determined by TEM [26].

In order to acquire additional insight on the crystal structure of the $BaTiO_3$ nanocrystals [30] XRD experiments using synchrotron radiation were performed and the data were analyzed by Rietveld refinement[41] and by using the atomic pair distribution function technique [46]. It was concluded that the three-dimensional atomic ordering in $BaTiO_3$ nanoparticles is more complex than in the bulk: it is cubic-like on average, but it shows locally slight distortions of a tetragonal-type. A description of these advanced crystallographic techniques and how these results were obtained overcome the scope of this chapter and we invite the reader to consult Ref. [30] for more information.

The second example shows how to use Debye scattering equation calculations to determine the shape and the preferential direction of growth of small and slightly elongated particles. In this case HfO_2 nanocrystals synthesized in benzyl alcohol are studied [33]. They exhibit an elongated shape as observed in HRTEM images (Figure 7.8a). This nanoparticle is a monocrystal and shows several lattice planes. For this reason it is a perfect example to be used for structure determination. The power spectrum (Figure 7.8b) is characterized by several sharp spots from which it is possible to extract the lattice distances and the angles between the planes. In this case it can be attributed to the monoclinic HfO_2 structure (JCPDS [43-1017] and ICSD 27313) oriented along the $[0\bar{1}1]$ direction as indexed in Figure 7.8b. The particle is elongated along the [100] direction. The monoclinic HfO_2 structure is characterized by similar lattice parameters: $a = 5.116$, $b = 5.172$ and $c = 5.295$ Å and $\beta = 99.2°$. Because the global precision in the determination of the dis-

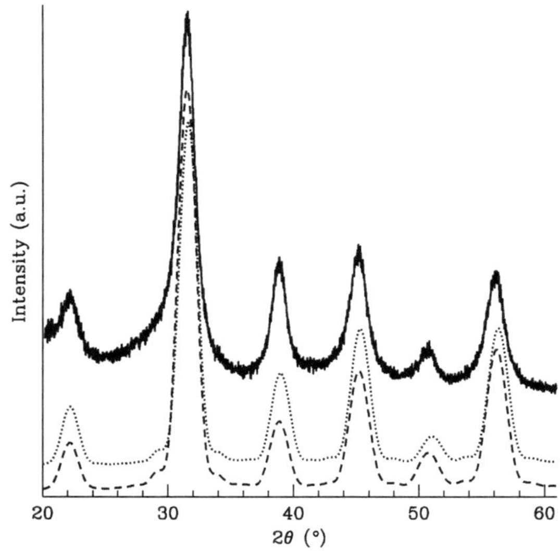

Fig. 7.7 XRD pattern of 6 nm $BaTiO_3$ nanoparticles synthesized by the benzyl alcohol route (*solid line*). Debye equation calculated for a 6 nm spherical $BaTiO_3$ nanoparticle with the cubic (*dotted line*) and tetragonal (*dashed line*) perovskite structure

tances by HRTEM is around 1–2% at most, in this particular case the particle could also be oriented along the [1$\bar{1}$0] with the long axis parallel to the [001] direction instead of [100]. To discriminate between the two possible growth directions experimental XRD patterns were compared to calculated ones.

The X-ray powder diffraction pattern of HfO_2 nanocrystals synthesized at 250 °C (Figure 7.9, solid line) shows broad diffraction peaks pointing to small crystallite sizes. Despite this broadening, the theoretical positions of the diffractions of the monoclinic HfO_2 structure match well with the experimental ones. To prove that the Debye scattering equation is a powerful tool to characterize the preferential growth direction of these nanocrystals several calculations for model particles were performed.

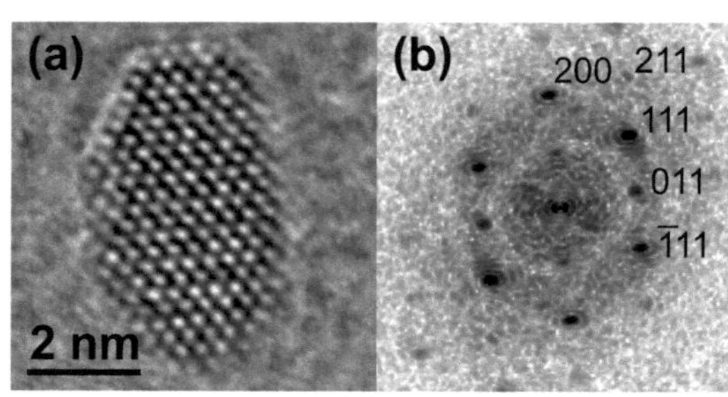

Fig. 7.8 a,b HRTEM image of a single HfO_2 particle **a** and its power spectrum **b**

Although the pattern calculated for a spherical particle of 4 nm in diameter (Figure 7.9, dotted line) fits relatively well the experimental data, the reflection at about $b = 0.393$ Å^{-1}, characteristic for the diffraction by the (002), (020), and (200) planes, is not well reproduced. These reflections are

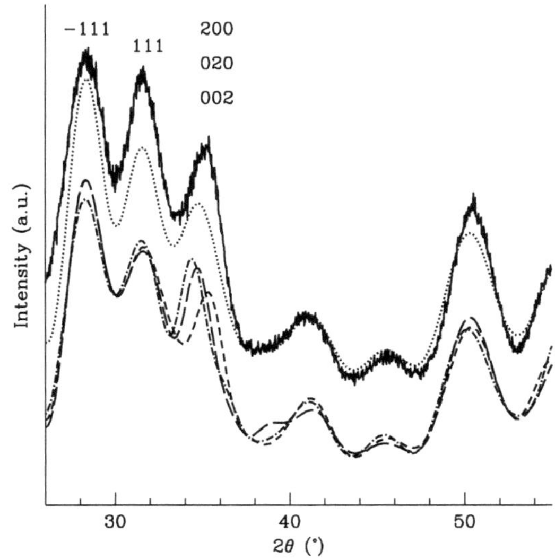

Fig. 7.9 Experimental pattern of HfO_2 nanoparticles synthesized at 250° C (*solid line*), calculated patterns for 4 nm spherical particles (*dotted line*), ellipsoidal particles of 6×3 nm elongated along the [100] (*short dashed line*), [010] (*long dashed line*), [001] (*dot, short dashed line*), respectively

superimposed in the experimental patterns, due to the broadening and because the lattice parameters of the HfO_2 monoclinic structure are similar. In the experimental patterns this peak is sharper, more intense and slightly shifted towards larger diffraction angles compared to the calculated ones, suggesting that the particles exhibit an elongated shape in one of these directions. It is reasonable to assume that the intensity as well as the shift of the experimental peak around $b = 0.393$ Å^{-1} is mainly caused by the diffraction of the (200) planes, and that the particles are elongated along the [100] axis. To prove this hypothesis, the Debye scattering equation was calculated for ellipsoidal particles (long radius = 3 nm and short radius = 1.5 nm) with the long axis either parallel to the [100], [010], or [001] directions, respectively (Figure 7.9 short dashed line, long dashed line and dot, short dashed line, respectively). All three calculated patterns seem to be quite similar in terms of peak intensity, the main difference lies in the position of the 002, 020 and 200 reflections. A comparison of the calculated and the experimental patterns clearly shows that the best agreement is found, when the particles are elongated along the [100] direction. The calculated peak position at $b = 0.393$ Å^{-1} fits perfectly with the experimental reflection, whereas calculations for a particle elongated along the [010] and along the [001] give a peak position of $b = 0.388$ Å^{-1} and $b = 0.383$ Å^{-1}, respectively. In conclusion, according to Debye calculations, the particles are about 6 nm in length and about 3 nm in width and display a preferential growth in the [100] crystal direction.

7.4 Combination of a Range of Techniques

The characterization of nanoparticles of a known crystalline structure can be usually achieved using only HRTEM and powder diffraction techniques. However, in the case of an unknown structure several techniques should be combined for structure elucidation. Single crystal diffraction is the most common technique for solving the structure of large crystals, however, it can not be used in the case of nanoparticles due to their small size. Moreover, if the structure is known, refinement using powder diffraction data (e.g. Rietveld method) are not effective and the broad diffraction peaks observed in small nanoparticles further complicate the task. In this case several techniques are generally used and the different information acquired from each technique are compared in order to build a reasonable structural model. It should be pointed out that such an approach rarely lead to the precise atomic positions in the elementary crystal cell. In our work such an approach was needed for the elucidation, among others, of the structure of vanadium oxide nanorods (cf. Chapter 4.1.3) and hybrid organic-inorganic materials based on rare earth oxides (cf. Chapter 4.2) and will be detailed below.

7.4.1 Hollandite-Type Vanadium Oxyhydroxide Nanorods

The structural determination of vanadium oxide nanorods synthesized in benzyl alcohol is an interesting and challenging example requiring the combination of several experimental techniques and theoretical tools [8]. Composition, structure, and morphology were thoroughly analyzed by neutron and synchrotron powder X-ray diffraction as well as by different electron microscopy techniques (SEM, HRTEM, EDX, EELS and SAED). The data obtained point to a hollandite-type structure $VO_{1.52}(OH)_{0.77}$ which, unlike other vanadates, contains oxide ions in the channels along the c-axis, with hydrogen atoms attached to the edge-sharing oxygen atoms, forming OH groups. *Ab initio* density-functional calculations with a local spin density approximation including orbital potential (LSDA + U with an effective U value of 4 eV) have been employed to extract the electronic structure. These calculations propose, on the one hand, that the electronic conductivity is based on electron hopping between neighboring V^{3+} and V^{4+} sites, and, on the other hand, that the oxide ions in the channels act as electron donors, increasing the fraction of V^{3+} cations, and thus leading to self-doping. Experimental and simulated electron energy-loss spectroscopy data confirm both the presence of V^{4+} and the validity of the density-of-states calculation.

The initial and preliminary analysis of the crystal structure of the vanadium oxyhydroxide nanorods involved Rietveld refinement of the X-ray powder diffraction pattern. The reflections observed matched well with the XRD pattern reported in the ICDD PDF database (No. 00-500-1797), corresponding to $V_{8.18}O_{16}{\cdot}1.46\ H_2O$ in the space group I4/m. To get more detailed structural information, including possible hydrogen atoms, high resolution probes such as synchrotron X-ray and neutron diffraction were performed. The complementarity of X-ray and neutron powder diffraction techniques was essential for the success of the complete crystal structure determination of $VO_{1.52}(OH)_{0.77}$ (Figure 7.10). The scattering amplitudes of vanadium, oxygen, and hydrogen atoms are in the ratio of 23:8:1 for X-rays and roughly -0.4:5.8:-3.7 for neutron diffraction. Consequently, hydrogen is almost invisible by X-rays, whereas vanadium is almost invisible for neutrons. On the other hand, in the neutron diffraction experiment, hydrogen and oxygen atoms strongly scatter. The structure obtained by refinement of the X-ray and neutron diffraction studies is shown in Figure 7.10, viewed in perspective (a), along the c-axis (b), and along the a-axis (c). A highly distorted VO_6 octahedron constitutes the basic structural unit building up the vanadium oxide framework. A detailed description of the peculiarity of this structure can be found in the original publication [8]. Below we will focus on a problem that could be only solved by using DFT calculations.

Surprisingly, the structure contains oxide ions O^{2-} placed in the center of the channels (noted O3 in the Figure 7.10a). Moreover, they look like they are doubled. However, such a splitting is, in fact, artificial, since instead of (0, 0, 0.101) for this site, equally good refinement is obtained if O3 is placed in the (0, 0, 0) position and subjected to strong anisotropic thermal motions. The representation of these central oxygens as ellipsoids, which are strongly elongated along the c-axis, in fact suggests a very shallow potential for the O3 at the center of the 2×2 channel. However, this could also be a split position, in which O3 occupies the upper position in one channel and the lower position in a neighboring channel. To resolve which of these two possibilities is correct (i.e., more stable), DFT calculations were performed in a supercell created from the $1 \times 1 \times 2$ hollandite unit cells. By comparing the total energies of both configurations, it was found that the first solution, i.e., O3 placed in the (0, 0, 0) position and subjected to strong anisotropic thermal motions, lies about 26 mRy (0.354 eV) below the energy corresponding to the split position. The morphology of the $VO_{1.52}(OH)_{0.77}$ nanoparticles was examined by SEM and TEM. An overview SEM image at low magnification illustrates that the product is almost exclusively composed of discrete and uniform particles with an anisotropic, rod-like shape (Figure 7.11a). The pointed ends of the rods give them an ellipsoidal, rice-like appearance, which is also confirmed by the TEM image in Figure 7.11b. Their length ranges from 140 to 500 nm, with an average of 330 nm, and the width varies between 50 and 105 nm, with an average of 80 nm. The local crystallinity of a single nanorod was further elucidated by HRTEM, revealing well-developed lattice fringes that extend over the whole area displayed in Figure 7.11c. The power spectra

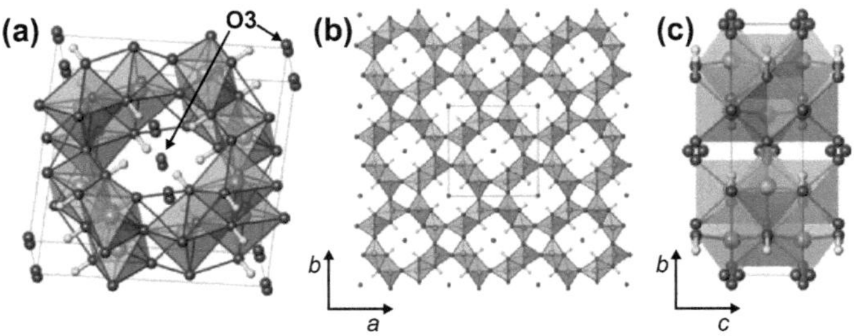

Fig. 7.10. **a–c** Crystal structure of $VO_{1.52}(OH)_{0.77}$ viewed **a** in perspective, **b** along the c-axis, and **c** along the a-axis. The vanadium atoms are represented by red, oxygen by blue, and hydrogen by green spheres

of the HRTEM images and SAED study point to single crystallinity of the nanorods with the main growth direction parallel to the [002].

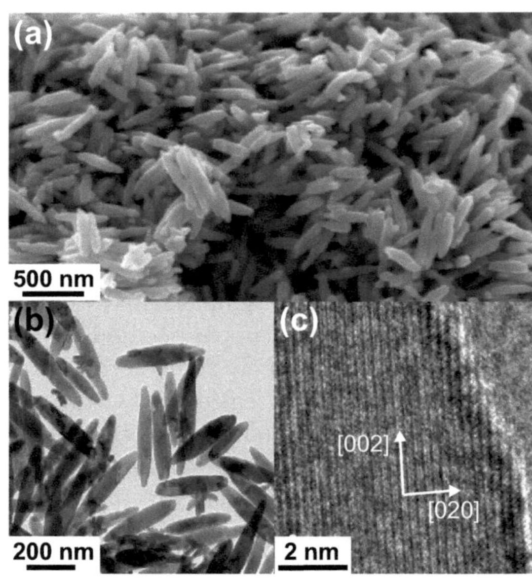

Fig. 7.11 **a** SEM and **b** TEM overview images of $VO_{1.52}(OH)_{0.77}$ nanorods. **c** HRTEM image of a part of a nanorod

Additional electrical and magnetic studies, and band structure calculations permitted to analyze and understand the peculiar physical properties of this particular system. However, they will not be detailed here as it would overcome the scope of this book. Finally, the $VO_{1.52}(OH)_{0.77}$ nanorods are an instructive example of how a simple, one-pot synthesis procedure can result in the formation of a highly complex material. The application of several state-of-the art characterization tools was required to elucidate the structural and physical properties. Some of the results were rather unexpected, and only the combination of various experimental techniques and theoretical modeling made it possible to understand and interpret them.

7.4.2 Lanthanide-Based Organic-Inorganic Hybrid Nanostructures

In this section the approach needed for the elucidation of the structure of hybrid organic-inorganic materials based on rare earth oxides introduced in Chapter 4.2 will be illustrated. The particular features of the ordered lamellar structure was revealed using several characterization techniques ranging from powder X-ray and electron diffractions, vibrational spectroscopy, solid state NMR, elemental analysis to emission spectroscopy. It turned out that the oxide layers are composed of only 3 layers (~0.6 nm) of crystalline rare earth oxide alternated with a double layer of benzoate or biphenolate species bridged to the oxide nanosheets (cf. Figure 4.11). Moreover, a detailed analysis of the reaction mixture gave valuable information on the formation mechanism of the hybrid structure and a valid explanation of the oxidation of the alcohol, used as solvent, to carboxylate species (cf. Chapter 5.5).

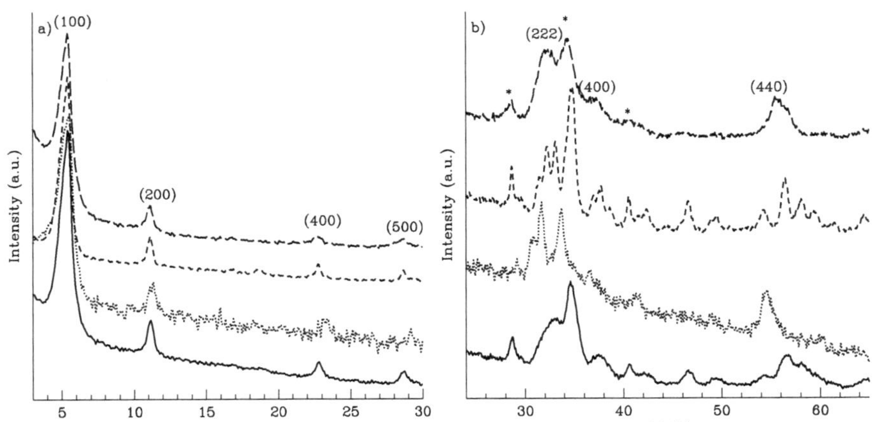

Fig. 7.12. a,b XRD (2θ – for the cobalt radiation) patterns of the benzoate based hybrid materials: yttrium (*solid line*), neodymium (*dotted line*), samarium (*dashed line*) (synthesized at 300 °C) and gadolinium (*long dashed line*) based hybrid materials. **a** Small angle region, **b** wide angle region in which the reflections due to the mesostructure are indicated by a star

Powder XRD provided complex patterns containing information at low diffraction angle about the mesostructural order (i.e. the lamellar periodicity) and at larger angle about the crystal structure of the RE_2O_3 type oxides

(Figure 7.12). The first most intense peak observed for a diffraction angle of around 5.5° corresponds to a distance of 1.8 nm, hence to the 100 reflection of the lamellar structure. The higher orders up to the fifth could be clearly assigned on the same figure. Even higher orders are visible on the wide angle part of the pattern (Figure 7.12b) especially the sixth at $2\theta \sim 35°$ which is as intense and overlaps with the near reflections due to the crystalline inorganic layers (zone $30° < 2\theta < 40°$). The extinction of the third order is typical for lamellar structures where the thickness of one layer is the double of the other. This is exactly what happens in our system where the inorganic layer is twice as thinner as the organic one. In fact, quantitative analysis of the XRD data performed following the approach of Ruland and Smarsly [9, 42] by fitting the experimental data with analytical expressions for the one-dimensional density function show that the long period deduced by this method is 1.75 nm and the thicknesses of the two layers are about 1.2–1.25 nm and 0.55–0.6 nm, respectively, and their variation in thickness is smaller than 0.05 nm [34]. The high number of reflections, due to the lamellar periodicity, denotes the high order and the monodispersity in term of organic and inorganic layer thicknesses. Surprisingly, the position of these reflections is the same for each sample. Hence, the thickness of the inorganic and organic layers is constant and does not vary regarding to the nature of the oxide.

Information about the crystal structure of the RE_2O_3 was impossible to extract from XRD data due to i) the sub-nanometric thickness of the layers giving rise to extremely broad diffraction peaks in the region $30° < 2\theta < 40°$ and ii) the overlapping with some reflections due to the lamellar periodicity (Figure 7.12b). HRTEM studies could not be performed due to the instability of the material under the electron beam due to the organic-inorganic structure. Additional structural data could be obtained by SAED under the TEM, but they could give a clear answer in one case only due to the preferential orientation of the particles on the TEM grid [16]. Hence, additional techniques should be used in order to get additional insight on the inorganic and organic parts forming the hybrid structure. Yttrium observation by ^{89}Y solid-state NMR (SSNMR) is usually a difficult task, owing to its long longitudinal relaxation time, low sensitivity and low gyromagnetic constant (γ). However ^{89}Y SSNMR spectra can be obtained by cross-polarization (CP) using the pool of ^{1}H spins [20]. The ^{89}Y cross polarization magic angle spinning (CPMAS) NMR spectra (Figure 7.13a and b) of two hybrid materials synthesized in 4-biphenylmethanol and benzyl alcohol are very similar, exhibiting two ^{89}Y resonances almost at the same isotropic chemical shift ($\delta_{iso} \sim 195$ and 215 ppm) [15]. The presence of two distinct ^{89}Y resonances at the same frequency for both materials indicates that the yttrium local environment and the crystal structures are analogous. The peaks observed in Figure 7.13a and b indicate the existence of two distinct yttrium environments. Additionally, the relative intensity of both peaks does not change significantly by performing various ^{89}Y CPMAS experiments at different contact times. This fact suggests that the ^{89}Y resonance at ca. 195 ppm correspond to a site more populated than that giving the peak at 215 ppm. The deconvolution of the ^{89}Y spectrum of the yttrium-benzoate compound yields a 2:1 ratio. This fact supports the hypothesis that the inorganic sheet is made up of three layers of yttrium oxide: one inner layer (which is not in contact with the organic molecules) and two outer layers of yttrium oxide. These findings supports the TEM and XRD measurements. Indeed, a thickness of 0.6 nm of Y_2O_3 would correspond to three yttrium layers. Thus, following this assumption, the ^{89}Y environments corresponding to the outer and inner layers should resonate at ca. 195 and ca. 215 ppm, respectively.

A confirmation of the existence of two distinct RE environments was also demonstrated by emission spectroscopy studies for europium doped yttrium and gadolinium-based materials. Indeed, europium is considered a powerful probe of the local environment because the europium emission lines are very sensitive to its local environment. High resolution spectra showed that two distinct environments exist, which can be assigned to a purely oxidic one (Eu ions in the inner layer) and a carboxylate bridged-like complex (Eu in the outer layer coordinated to carboxylate species as well) [17, 44].

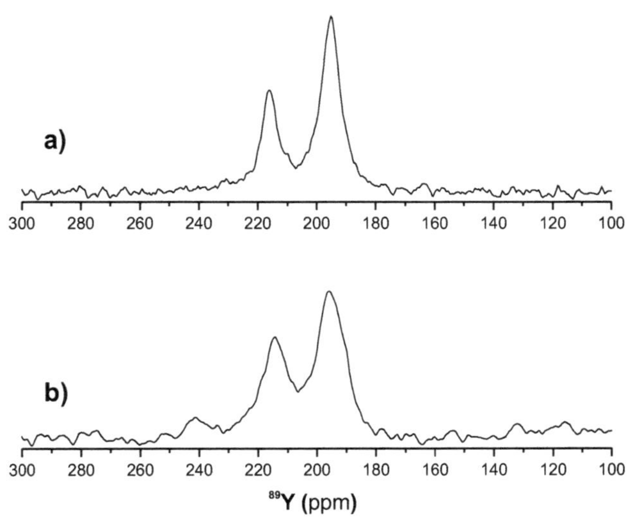

Fig. 7.13. ^{89}Y CPMAS spectra of **a** yttrium-biphenolate and **b** yttrium-benzoate [15]

In order to study the composition of the organic layer forming the hybrid materials the samples were extensively studied by vibrational spectroscopy and ^{1}H and ^{13}C SSNMR. The powder of the yttrium based hybrid materials synthesized in benzyl alcohol and 4-biphenylmethanol were analyzed by ^{1}H-broadband decoupled ^{13}C CPMAS SSNMR. Figure 7.14 show a typical example for the materials synthesized in 4-biphenylmethanol. Aside from the anisotropic sidebands, only three peaks were observed. All of them, namely the signal of the carboxylate group in the range of 171–176 ppm, the aromatic carbon signals at 126–133 ppm and the carbons bridging the phenyl rings at 140.5–142.6 can be attributed to biphenolate species. The absence of the methylene group from the starting material (4-biphenylmethanol), which resonates at ca. 64 ppm, indicates its complete conversion to a biphenolate species. In order to support SSNMR studies and prove that similar behavior

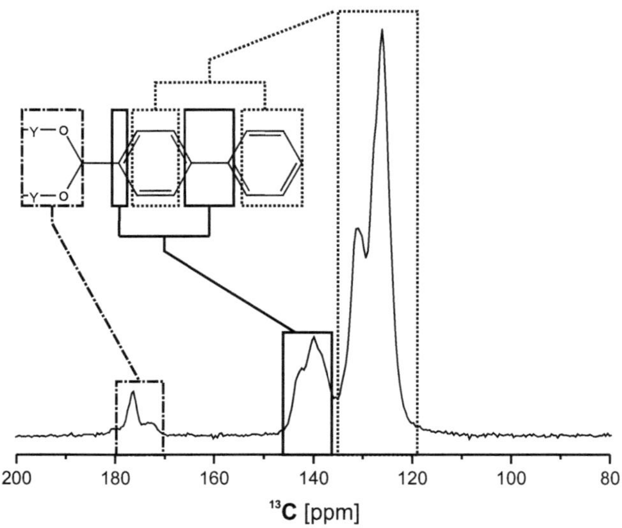

Fig. 7.14. ^{13}C CPMAS NMR spectrum of the yttrium based biphenolate hybrid material

is also obtained with the other RE ions[1], vibrational spectra (FT-IR and Raman) were recorded. The vibrational spectra of metal-organic complexes and salts of carboxylic acids have been studied extensively in the literature [22]. The comparison of vibrational data with X-ray crystal structures provides information about the kind of bond of the carboxylic group to the metal center, just by calculating the difference of the $\nu_a(CO_2)$ and $\nu_s(CO_2)$ stretching frequencies [5]. In the IR spectra of the hybrid materials synthesized in benzyl alcohol and in 4-biphenylmethanol, $\nu_a(CO_2)$ are located around 1570 cm^{-1} and $\nu_s(CO_2)$ around 1390 cm^{-1}, giving a difference of $\Delta \simeq 180$ cm^{-1} [34, 17, 15]. These values agree well to a bridged kind of bonding. This supports solid state NMR findings and further proves that the organic moieties and the bonding types are independent of the rare earth element.

In conclusion, the combination of various characterization techniques allowed to have a precise picture of the peculiar features of these hybrid materials and draw a structural model (cf. Figure 4.11). It is worth pointing

[1] Indeed, similar SSNMR studies can not be performed in the presence of paramagnetic ions such as the other rare earth used (e.g. Nd, Sm, Eu, Gd, Tb, Er)

out that the structure could have been solved in a much easier way if single crystals of several micrometers in size could have been obtained. Combining different characterization techniques for the elucidation of structures is a more and more common approach especially in the field of nanostructured and hybrid materials.

References

1. Batson P.E.and Dellby, N.K.O.: Sub-angstrom resolution using aberation corrected electron optics. Nature **418**, 617–620 (2002)
2. Chiche, D., Digne, M., Revel, R., Chaneac, C., Jolivet, J.P.: Accurate determination of oxide nanoparticle size and shape based on X-ray powder pattern simulation: Application to boehmite AlOOH. J. Phys. Chem. C **112**, 8524–8533 (2008)
3. Coelfen, H., Pauck, T.: Determination of particle size distributions with angstrom resolution. Colloid Polym. Sci. **275**, 175–180 (1997)
4. Cowley, J.M., Moodie, A.F.: The scattering of electrons by atoms and crystals. I. A new theoretical approach. Acta Cryst. **10**, 609–619 (1957)
5. Deacon, G.B., Phillips, R.J.: Relationships between the carbon-oxygen stretching frequencies of carboxylate complexes and the type of carboxylate coordination. Coord. Chem. Rev. **33**, 227–250 (1980)
6. Debye, P.: Zerstreuung von Röntgenstrahlen. Ann. Phys. **351**, 809–823 (1915)
7. Djerdj, I., Arcon, D., Jaglicic, Z., Niederberger, M.: Nonaqueous synthesis of manganese oxide nanoparticles, structural characterization, and magnetic properties. J. Phys. Chem. C **111**, 3614–3623 (2007)
8. Djerdj, I., Sheptyakov, D., Gozzo, F., Arcon, D., Nesper, R., Niederberger, M.: Oxygen self-doping in hollandite-type vanadium oxyhydroxide nanorods. J. Am. Chem. Soc. **130**, 11,364–11,375 (2008)
9. Garnweitner, G., Smarsly, B., Assink, R., Ruland, W., Bond, E., Brinker, C.J.: Self-assembly of an environmentally responsive polymer/silica nanocomposite. J. Am. Chem. Soc. **125**, 5626–5627 (2003)
10. Gnutzmann, V., Vogel, W.: Structural sensitivity of the standard platinum/silica catalyst EuroPt-1 to hydrogen and oxygen exposure by in situ X-ray diffraction. J. Phys. Chem. **94**, 4991–4997 (1990)
11. Guinier, A.: X-ray diffraction in crystals, imperfect crystals, and amorphous bodies. Dover, New York (1994)
12. Hall, B.: Debye function analysis of structure in diffraction from nanometer-sized particles. J. Appl. Phys. **87**, 1666–1675 (2000)
13. Hall, B., Monot, R.: Calculating the Debye-Scherrer diffraction pattern for large clusters. Comp. in Phys. **5**, 414–417 (1991)
14. Hall, B.D., Ugarte, D., Reinhard, D., Monot, R.: Calculations of the dynamic Debye-Scherrer diffraction patterns for small metal particles. J. Chem. Phys. **103**, 2384–2394 (1995)
15. Karmaoui, M., Mafra, L., Sá Ferreira, R.A., Rocha, J., Carlos, L.D., Pinna, N.: Photoluminescent rare-earth based biphenolate lamellar nanostructures. J. Phys. Chem. C **111**, 2539–2544 (2007)
16. Karmaoui, M., Sá Ferreira, R.A., Carlos, L.D., Pinna, N.: Lanthanide-based lamellar nanohybrids: The case of erbium. Mater. Sci. Eng. C **27**, 1368–1371 (2007)

17. Karmaoui, M., Sá Ferreira, R.A., Mane, A.T., Carlos, L.D., Pinna, N.: Lanthanide-based lamellar nanohybrids: Synthesis, structural characterization, and optical properties. Chem. Mater. **18**, 4493–4499 (2006)
18. Keys, R.J., Garratt-Reed, A.J., Goodhew, P.J., Lorimer, G.W.: Introduction to scanning transmission electron microscopy. Bios Scientific Publishers, Oxford (1997)
19. Liu, J.: Advanced electron microscopy characterization of nanostructured heterogeneous catalysts. Microsc. Microanal. **10**, 55–76 (2004)
20. Merwin, L.H., Sebald, A.: The first yttrium-89 CP-MAS spectra. J. Magn. Res. **88**, 167–171 (1990)
21. Meurig Thomas, J., Midgley, P.A.: High-resolution transmission electron microscopy: The ultimate nanoanalytical technique. Chem. Commun. pp. 1253–1267 (2004)
22. Nakamoto, K.: Infrared and Raman spectra of inorganic and coordination compounds part B. Wiley, New York (1997)
23. Neri, G., Bonavita, A., Micali G. Rizzo, G., Pinna, N., Niederberger, M., Ba, J.: Effect of the chemical composition on the sensing properties of In_2O_3-SnO_2 nanoparticles synthesized by a non-aqueous method. Sens. Actuators B **130**, 222–230 (2008)
24. Neri, G., Bonavita, A., Rizzo, G., Galvagno, S., Pinna, N., Niederberger, M., Capone, S., Siciliano, P.: Towards enhanced performances in gas sensing: SnO_2 based nanocrystalline oxides application. Sens. Actuators B **122**, 564–571 (2007)
25. Niederberger, M., Bartl, M.H., Stucky, G.D.: Benzyl alcohol and titanium tetrachloride: A versatile reaction system for the nonaqueous and low-temperature preparation of crystalline and luminescent titania nanoparticles. Chem. Mater. **14**, 4364–4370 (2002)
26. Niederberger, M., Pinna, N., Polleux, J., Antonietti, M.: A general soft chemistry route to perovskites and related materials: Synthesis of $BaTiO_3$, $BaZrO_3$ and $LiNbO_3$ nanoparticles. Angew. Chem. Int. Ed. **43**, 2270–2273 (2004)
27. O'Keefe, M., Hetherington, C., Wang, Y., Nelson, E., Turner, J., Kisielowski, C., Malm, J.O., Mueller, R., Ringnalda, J., Pan, M., Thust, A.: Sub-angstrom high-resolution transmission electron microscopy at 300 keV. Ultramicroscopy **89**, 215–241 (2001)
28. Ould-Ely, T., Prieto-Centurion, D., Kumar, A., Guo, W., Knowles, W., Asokan, S., Wong, M., Rusakova, I., Luttge, A., Whitmire, K.: Manganese(II) oxide nanohexapods: Insight into controlling the form of nanocrystals. Chem. Mater. **18**, 1821–1829 (2006)
29. Patterson, A.L.: The Scherrer formula for X-ray particle size determination. Phys. Rev. **56**, 978–982 (1939)
30. Petkov, V., Gateshki, M., Niederberger, M., Ren, Y.: Atomic-scale structure of nanocrystalline $Ba_xSr_{1-x}TiO_3$ (x=1, 0.5, 0) by X-ray diffraction and the atomic pair distribution function technique. Chem. Mater. **18**, 814–821 (2006)
31. Pinna, N.: X-ray diffraction from nanocrystals. Progr. Colloid Polym. Sci. **130**, 29–32 (2005)
32. Pinna, N., Antonietti, M., Niederberger, M.: A novel nonaqueous route to V_2O_3 and Nb_2O_5 nanocrystals. Colloids Surf., A **250**, 211–213 (2004)
33. Pinna, N., Garnweitner, G., Antonietti, M., Niederberger, M.: Non-aqueous synthesis of high-purity metal oxide nanopowders using an ether elimination process. Adv. Mater. **16**, 2196–2200 (2004)
34. Pinna, N., Garnweitner, G., Beato, P., Niederberger, M., Antonietti, M.: Synthesis of yttria-based crystalline and lamellar nanostructures and their formation mechanism. Small **1**, 112–121 (2005)
35. Pinna, N., Grancharov, S., Beato, P., Bonville, P., Antonietti, M., Niederberger, M.: Magnetite nanocrystals: Nonaqueous synthesis, characterization, and solubility. Chem. Mater. **17**, 3044–3049 (2005)

36. Pinna, N., Neri, G., Antonietti, M., Niederberger, M.: Nonaqueous synthesis of nanocrystalline semiconducting metal oxides for gas sensing. Angew. Chem. Int. Ed. **43**, 4345–4349 (2004)
37. Pinna, N., Wild, U., Urban, J., Schlogl, R.: Divanadium pentoxide nanorods. Adv. Mater. **15**, 329–331 (2003)
38. Polleux, J., Pinna, N., Antonietti, M., Hess, C., Wild, U., Schlogl, R., Niederberger, M.: Ligand functionality as a versatile tool to control the assembly behavior of preformed titania nanocrystals. Chem. Eur. J. **11**, 3541–3551 (2005)
39. Polleux, J., Pinna, N., Antonietti, M., Niederberger, M.: Ligand-directed assembly of preformed titania nanocrystals into highly anisotropic nanostructures. Adv. Mater. **16**, 436–439 (2004)
40. Polleux, J., Pinna, N., Antonietti, M., Niederberger, M.: Growth and assembly of crystalline tungsten oxide nanostructures assisted by bioligation. J. Am. Chem. Soc. **127**, 15,595–15,601 (2005)
41. Rietveld, H.M.: A profile refinement method for nuclear and magnetic structures. J. Appl. Crystallogr. **2**, 65–71 (1969)
42. Ruland, W., Smarsly, B.: SAXS of self-assembled nanocomposite films with oriented two-dimensional cylinder arrays: An advanced method of evaluation. J. Appl. Crystallogr. **38**, 78–86 (2005)
43. Ruska, E.: The development of the electron microscope and of electron microscopy. Rev. Mod. Phys. **59**, 627–638 (1987)
44. Sá Ferreira, R.A., Karmaoui, M., Nobre, S.S., Carlos, L.D., Pinna, N.: Optical properties of lanthanide-doped lamellar nanohybrids. Chem. Phys. Chem. **7**, 2215–2222 (2006)
45. Scherrer, P.: Göttinger Nachrichten (1918)
46. Toby, B.H., Egami, T.: Accuracy of pair distribution function analysis applied to crystalline and non-crystalline materials. Acta Cryst. A **48**, 336–346 (1992)
47. Urban, J.: Crystallography of clusters. Cryst. Res. Technol. **33**, 1009–1024 (1998)
48. Van Dyck, D.: Electron microscopy: Principles and fundamentals, chap. High-resolution electron microscopy, pp. 109–162. Wiley-VCH (1997)
49. Vogel, W.: X-ray diffraction from clusters. Cryst. Res. Technol. **33**, 1141–1154 (1998)
50. Wang, Z.: Characterization of nanophase materials. Wiley-VCH, Weinheim (1999)
51. Wang, Z.L.: Transmission electron microscopy of shape-controlled nanocrystals and their assemblies. J. Phys. Chem. B **104**, 1153–1175 (2000)
52. Wang, Z.L.: New developments in TEM for nanotechnology. Adv. Mater. **15**, 1497–1514 (2003)
53. Warren, B.: X-ray diffraction. Dover, New York (1990)
54. Zanchet, D., Hall, B.D., Ugarte, D.: Structure population in thiol-passivated gold nanoparticles. J. Phys. Chem. B **104**, 11,013–11,018 (2000)
55. Zhong, X., Xie, R., Sun, L., Lieberwirth, I., Knoll, W.: Synthesis of dumbbell-shaped manganese oxide nanocrystals. J. Phys. Chem. B **110**, 2–4 (2006)
56. Zitoun, D., Pinna, N., Frolet, N., Belin, C.: Single crystal manganese oxide multipods by oriented attachment. J. Am. Chem. Soc. **127**, 15,034–15,035 (2005)

Chapter 8
Properties and Applications

8.1 Introduction

The unique characteristics of transition metal oxides make them the most diverse class of materials, with properties covering almost all aspects of materials science and solid state physics [82]. Metal oxides experience the same trend as many other advanced materials: The miniaturization of functional devices in emerging technologies such as gas sensing, catalysis, energy storage and conversion and electroceramics demands for the production of these materials with the highest possible purity, small crystallite size, narrow particle size distributions and well-defined particle morphology and chemical composition. The high scientific as well as technological interest in nanoparticles lies in the fact that the dependence of the chemical and physical properties of nanocrystalline solids on particle size and shape provides a powerful tool to tailor the properties of a material.

In this chapter we shall introduce some physical properties of selected metal oxide nanoparticles synthesized in organic solvents. The chapter is organized in the following sections: i) 8.2 magnetic properties of intrinsic magnetic oxides (e.g. ferrites) and oxides doped with paramagnetic ions (e.g. manganese doped zinc oxide), ii) 8.3 photoluminescence properties of rare earth and semiconductor metal oxides, iii) 8.4 (photo)catalytic properties, iv) 8.5 gas sensing devices fabricated using metal oxide nanoparticles as active component for the detection of various gases and v) 8.6 the possible use of metal oxide nanoparticles in medical applications such as magnetic resonance imaging and cancer treatment.

8.2 Magnetic Properties

In this section we will concentrate on the magnetic properties of oxide nanoparticles made of paramagnetic transition metal atoms (e.g. iron oxides and ferrites) and doped semiconductor oxide nanoparticles (e.g. ZnO:Co).

8.2.1 Magnetic Metal Oxides

Iron oxides and ferrites such as γ-Fe_2O_3, Fe_3O_4, $CoFe_2O_3$ have a net magnetic moment due to the antiparallel arrangement of the moment of the atoms in the octahedral and tetrahedral sites of the inverse spinel structure (e.g. Fe^{2+}, Fe^{3+}, Co^{2+}). Due to their small size (usually <30–40 nm) magnetic oxide nanoparticles are generally single-domain [24] and show a superparamagnetic behavior depending on the temperature, their size and their composition. The magnetization of ferromagnetic and ferrimagnetic nanoparticles above a given temperature, called blocking temperature (T_B), displays no hysteresis as a function of increasing and decreasing magnetic field (M-H). This is caused by the thermal energy kT that, in the superparamagnetic regime, is larger than the energy required to change the direction of the magnetic moment of a particle. Therefore, above T_B the nanoparticles do not retain any magnetization in the absence of an externally applied magnetic field.

The T_B increases with the number of magnetic atoms and thus, with the size of the nanoparticles. Therefore, its precise evaluation requires homogeneous samples characterized by a low size distribution. The group of T. Hyeon reported the synthesis of monodisperse iron oxide nanoparticles with a continuous size spectrum of 4–15 nm (cf. Chapter 3.6) [72]. In the same article the authors reported the magnetic properties and the T_B in function of the size (Figure 8.1), evaluated by comparing magnetization versus increasing temperature (M-T) of the zero-field-cooled (ZFC) and field-cooled (FC). T_B varies from 14 K to 275 K when the iron oxide particle size increases from 4 to 15 nm.

The total magnetic moment of a magnetic single-domain nanoparticle (i.e. the magnetization at saturation extracted from M-H experiments) should be, in principle, directly proportional to the number of magnetic atoms in the particle. However, the magnetization at saturation measured for nanoparticles is often lower than expected. This is principally due to two reasons: i) the reduction of the particle size results in a large amount of magnetic atoms on the surface of the particle. These atoms have fewer nearest neighbors to which they can interact. Therefore, it is more likely that their moment become randomly oriented at the surface (spin-canting) and that higher applied fields are needed to align them parallel to it. ii) Moreover, when the syntheses are performed in the presence of surfactants a large amount of organic species are present at the particle surface and can contribute to 30% or more of the

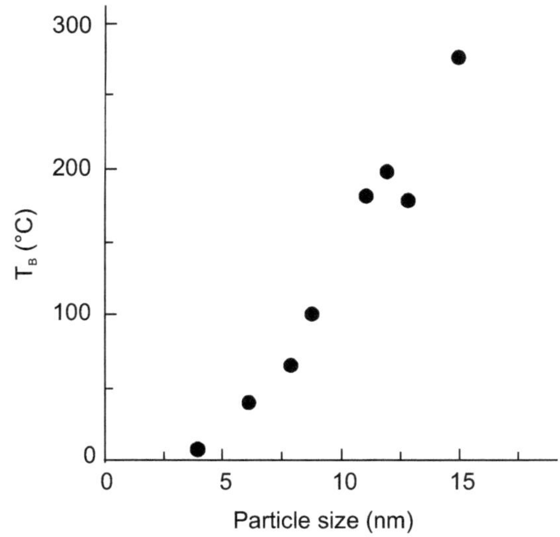

Fig. 8.1 Size dependence of the blocking temperature of maghemite nanoparticles, T_B, deduced from M-T experiments in function of the particle size. Experimental points from Ref. [72]

total weight of the sample [83]. Therefore, the magnetization is artificially lowered by the weight fraction of these impurities.

Surfactant-free routes have also proven to be able to produce high quality iron oxide nanocrystals (cf. Table 4.1). Syntheses performed in benzyl alcohol from iron acetylacetonate lead to magnetite nanoparticles with tunable size from 12 to 25 nm [77]. The magnetization at saturation is comparable to the one of the bulk magnetite (Figure 8.2), highlighting the high quality and purity of the as-synthesized nanocrystals. Indeed, from HRTEM studies, they were found single crystalline in nature with no detectable core defects (cf. Figure 7.2c). The blocking temperature lies close or above room temperature for the 12 and 25 nm sized nanoparticles, respectively.

8.2.2 Diluted Magnetic Semiconductors

These materials are typically III-V or II-VI semiconductors doped with transition metals [32]. In the past five years, the revival of the field mainly occurs due to the assumption that room temperature ferromagnetism is promoted by the semiconductor charge carriers, based on the so-called Zener model adapted to DMS [21]. However, many of the claims of high-temperature ferromagnetism in DMS systems are controversial owing to the tendency of these materials to phase-separate into a semiconductor matrix laced with ferromagnetic metal or oxide domains. Consequently, the main point of any new study relies on the exclusion of any secondary phase as the origin for ferromagnetism. This point is critical for nanoscale systems as they tend to expel the doping atoms from the nanocrystal's core.

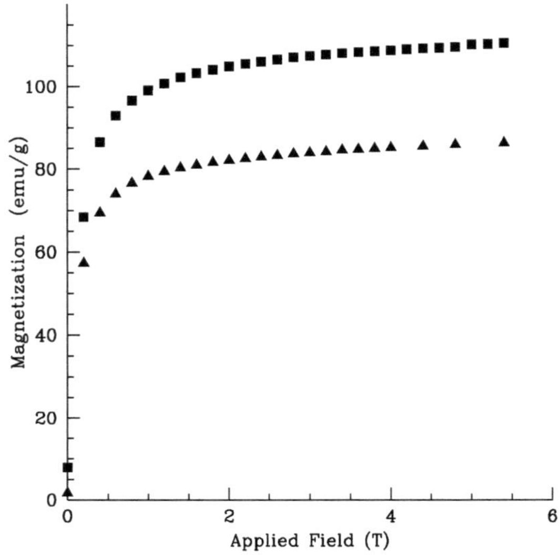

Fig. 8.2 Magnetization curves at 5 K of two samples of magnetite nanoparticles synthesized in benzyl alcohol having average sizes of 12 (▲) and 25 nm (■), respectively [77]

Recently, the quest for high temperature ferromagnetic semiconductors was dominated by theoretical and experimental studies on zinc oxide and gallium nitride. These materials gave rise to further investigations on oxides such as TiO_2, ZrO_2 and HfO_2. Interestingly, doped oxides were rarely considered as potential DMS. However, as recently predicted by Ostanin et al. [69], doped ZrO_2 should be considered as one of the few suitable materials.

The exact origin of the magnetism of DMS is still unclear. However, for homogeneous systems it seems to be strongly dependent on dopant/matrix interaction, carrier concentration, defects, and thus on synthetic conditions. For this reason meticulous characterizations of the doped nanoparticles obtained, with emphasis on the homogeneity and local environment of the magnetic ions diluted in the matrix, are needed.

Cobalt and manganese doped zinc oxide nanowires were synthesized in trioctylamine at 310°C [15, 113]. The synthesis approach led to bundles of nanowires that are tethered to a common hexagonal base. The diameter of the nanowires is statistically invariant regardless of the dopant concentration and the average nanowire diameter and length are approximately 35 nm and 3–4 μm, respectively. The metal dopant concentration could be varied from 1% to 15% and was found to be homogeneous and no evidence of a lateral or longitudinal concentration gradient in individual nanostructures was detected. ZnO nanowires doped with 2.1% of cobalt exhibit magnetic hysteresis with coercive fields decreasing from 230 Gauss at 2 K to 140 Gauss at 300 K. This is the usual observed behavior for ferromagnetic materials. The magnetization does not saturate and reaches the value of 3.5 μB/Co atom at μ_0H = 5.0 T which corresponds to the experimental value for isolated Co^{2+} ions. Magnetization versus temperature (M–T) curves, at a constant applied field of μ_0H = 5.0 mT, remains at a non-zero value in the temperature range of 20–350 K, proving the existence of a high-Curie-temperature ferromagnetism. In the case of manganese, for all concentrations, the as-synthesized samples never show any deviation from a paramagnetic behavior.

The reaction of metal oxide precursors in benzyl alcohol also proved to be a versatile approach for the synthesis of transition metal doped ZnO [16, 23], TiO_2 [22] and ZrO_2 [17] (cf. Table 4.3).

Doped zinc oxide nanoparticles were synthesized by reacting zinc acetate with benzyl alcohol at 250°C [16]. This enables the incorporation of several dopants into the ZnO matrix, such as Mn, Fe and Co. Moreover, by varying the amount of benzyl alcohol with anisole, used as inert co-solvent, the morphology of the particles can be tuned. For example, Co-doped ZnO synthesized in benzyl alcohol/anisole (5/95%) yield small rods whereas particles are obtained for Mn-doped ZnO synthesized in pure benzyl alcohol. Surprisingly, the use of anisole affects the magnetic behavior as well. Mn-doped ZnO always displays a paramagnetic behavior while Co-doped particles are paramagnetic or ferromagnetic depending on the solvent used. The zero-field-cooled/field-cooled (ZFC-FC) measurements (Figure 8.3) show no deviation from Curie's law for nanoparticles synthesized in benzyl alcohol/anisole (5/95%) while for the synthesis in pure benzyl alcohol, the ZFC-FC curve deviates from a Curie law above 30 K. Above this temperature the magnetic moment shows a typical ferromagnetic behavior until room temperature similar to other reported Co-doped ZnO particles [15]. For both cobalt and manganese doped zinc oxide, independently of the synthetic condition, it was shown that the dopant substitutes onto zinc sites in the wurtzite lattice, with no detectable phase impurities or clustering. Hence, the observance of ferromagnetism appears to be clearly related to the core properties of cobalt doped ZnO nanocrystals. Co- and Mn-doped ZnO nanowires were also synthesized in benzyl alcohol from the corresponding metal acetylacetonates as precursors [23]. The Co-doped ZnO powders are ferromagnetic with a Curie temperature exceeding room temperature. Conversely, the Mn-doped samples show antiferromagnetic correlations with a possible transition to an antiferromagnetic ground state below T_N = 10 K. The results suggest that the magnetic properties are extremely sensitive to the type of dopant and are in agreement with previous studies.

The synthesis of doped zirconia led to high quality spherical nanocrystals of 3–4 nm [17]. The particles are monocrystalline in nature and do not present any core defects. Previous theoretical calculations on zirconia indicate that the ferromagnetic interactions are closely related to the oxidation state of the magnetic ion [69]. EPR experiments of the Mn-doped samples reveal the coexistence of manganese oxidation state (II) and (III), their ratio depending on the initial manganese concentration. The magnetic measurements show mainly a paramagnetic behavior. Nevertheless, for a larger Mn concentration, antiferromagnetic interactions dominate the ZFC-FC curves. As ferromagnetic behavior is expected upon increasing the concentration and

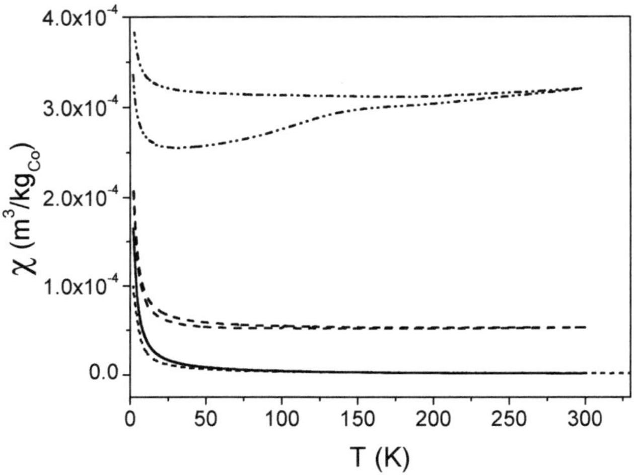

Fig. 8.3. ZFC-FC magnetization curves of Co-doped ZnO synthesized in benzyl alcohol/anisole (5/95%) at 0.81% (*full line*) and 4.09% (*dotted line*), Co-doped ZnO in pure benzyl alcohol at 0.77% (*dashed line*) and 3.88% (*dot-dot-dashed line*)

the oxidation state of the dopants [69], modification of the synthesis conditions and post-synthesis annealing might permit to obtain ferromagnetic interactions.

The reaction between either $TiCl_4$ or $Ti(O\textit{i}Pr)_4$ and benzyl alcohol or 2-butanone results in the formation of phase pure TiO_2-anatase nanoparticles. Doping was achieved by adding 3 mol% iron(III) and cobalt(II) acetylacetonates [22]. According to XRD, impurity peaks, which might be attributed to the presence of a crystalline by-product, were not detected, confirming the phase purity of the final products. However, the doping efficiency was generally low. The highest doping (0.7 mol%) was achieved for Co-doped TiO_2 synthesized from $Ti(O\textit{i}Pr)_4$. At this low doping no interaction between magnetic atoms was detected, i.e., the material present a typical paramagnetic response.

8.3 Photoluminescent Metal Oxides

8.3.1 Rare Earth-Based Nanostructures

In the recent literature the synthesis of rare earth (RE) metal oxide particles in organic solvents was reported and their photoluminescence properties measured (cf. Table 3.1) [92, 102, 70, 93]. Si et al. [92] reported a synthetic approach leading to almost every rare earth sesquioxide (general formula RE_2O_3). In a following work the same authors reported the fabrication of pure RE_2O_3 and Y_2O_3:Eu [93]. The reaction of yttrium carbonate and europium carbonate in oleic acid and TOPO leads to Y_2O_3:Eu nanoparticles and nanodisks [102]. Recently, Zhao et al. [117] reported a synthesis approach to lanthanide oxysulfides Eu_2O_2S and Gd_2O_2S:Eu. Finally, pure Eu_2O_3 and Tb_2O_3 nanoparticles were obtained by microwave assisted synthesis [70]. The RE metal oxide nanoparticles cited above were all synthesized in the presence of surfactants. Generally the study of the optical properties of these materials is limited to emission and excitation spectra and no details about other important properties such as emission quantum yield (QY), radiance and life time were given. In some rare cases QY up to 3-4% were reported [102, 117]. Those values cannot be compared to the one obtained for rare earth based phosphors (QY $>$ 90%), which are typically composed of micrometric sized particles. It is also worth mentioning that such QY cannot compete with the ones obtained for II-VI semiconductor quantum dots and other fluorescent molecules commonly used for biological labeling. This definitively limits the field of application of these RE metal oxide nanoparticles.

When the "benzyl alcohol route" is used for the fabrication of RE oxides, ordered organic-inorganic hybrid nanostructures, based on M_2O_3 type thin oxide layers (M= Y, Gd, Nd, Er, Sm) are obtained [45, 43, 44, 76, 73]. Their formation mechanism, morphological and structural properties were already described in detail in Chapters 4.2, 5.5 and 7.4.2, respectively. Briefly, it was demonstrated that the synthetic approach leads to crystalline lanthanide oxide layers regularly separated from each other by organic layers of intercalated benzoate or biphenolate molecules. The thickness of the crystalline nanosheets was ~0.6 nm, while the thickness of the organic layer was ~1.2 or ~1.8 nm, depending on the alcohol used (i.e. benzyl alcohol or 4-biphenylmethanol). Additionally, it was shown that the yttrium- and gadolinium-based nanohybrids may be doped with Eu^{3+}, Tb^{3+}, and Nd^{3+} luminescent lanthanide ions. The optical properties of several RE ions emitting in the IR and visible were studied. Erbium- and neodymium-based hybrid nanostructures synthesized in benzyl alcohol show interesting emission properties in the near infrared (NIR) when excited in the organic subphase. Figure 8.4a shows the NIR spectrum of the erbium-based material excited at 290 nm where only the phenyl rings of the benzoate molecules in between the layers absorb [44]. The emission lines are assigned to the Er^{3+} intra-$4f^{11}$ transition, ${}^4I_{13/2} \rightarrow {}^4I_{15/2}$. It should be pointed out that it is not usual to observe room temperature emission from Er^{3+}-based hybrid materials containing organic groups, particularly if hydroxyl groups are present, which are known to prevent Er^{3+} radiative emission. However, in this case the Er^{3+} ions are active at room temperature indicating that they are efficiently protected from non-radiative deactivations. Figure 8.4a also shows the excitation spectrum monitored within the Er^{3+} ${}^4I_{13/2} \rightarrow {}^4I_{15/2}$ transition. The spectrum is formed of a series of straight lines assigned to transitions from the ground state $^4I_{15/2}$ to intra-4f^{11} excited levels and of a band peaking around 290 nm, which can be ascribed to the phenyl rings of the benzoate molecules [44]. The presence of such a band indicates that phenyl rings of the benzoate molecules in between the layers absorb energy and then further transfer it to the Er^{3+} ions at room temperature.

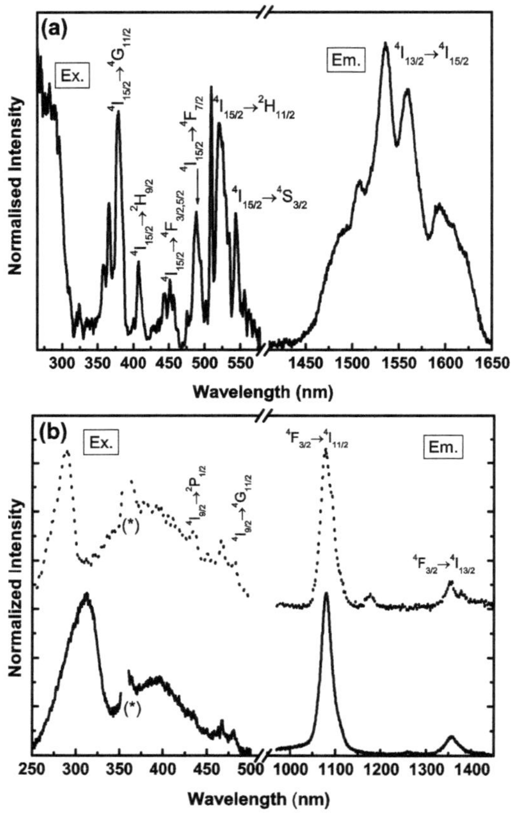

Fig. 8.4. **a** Room temperature NIR emission (Em.) and UV/VIS excitation (Ex.) spectra of the erbium-based hybrid material synthesized in benzyl alcohol excited at 290 nm and monitored at 1535 nm, respectively. **b** Room-temperature excitation (Ex.) and NIR emission (Em.) spectra of the yttrium- (*solid line*) and gadolinium-based (*dotted line*) hybrid materials synthesized in benzyl alcohol doped with Nd^{3+} monitored around 1060 nm and excited under 291-312 nm, respectively. (* denotes the detection of the monitoring wavelength harmonic).

Similarly, by exciting the yttrium and gadolinium-based hybrid materials doped with Nd^{3+} within the organic subphase emission lines in the NIR are observed. They are assigned to the $^4F_{3/2} \rightarrow {}^4I_{11/2,13/2}$ Nd^{3+} intra-4f^3 transitions. The Nd^{3+} ions are also active at room temperature. Figure 8.4b shows the excitation spectrum monitored within the Nd^{3+} NIR more intense transition. The spectrum is formed of two large broad bands peaking around 290-312 nm and around 400 nm. As in the previous example, these bands can be ascribed to the excited states of the phenyl rings of the benzoate molecules demonstrating that there is an active visible-to-NIR energy conversion channel at room temperature [86].

Yttrium- and gadolinium-based hybrid materials doped with Eu^{3+} and Tb^{3+} luminescent lanthanide ions present emissions in the red and green spectral regions. Figure 8.5a shows the room-temperature excitation spectra monitored within the $^7F_2$ manifold for the hybrid materials doped with Eu^{3+} ions. The spectra are formed of a broad band peaking at ca. 284 nm ascribed

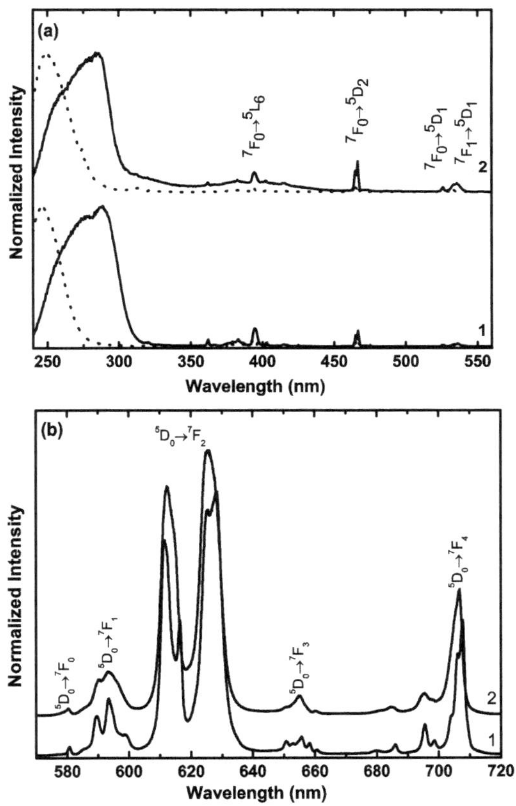

Fig. 8.5. a Room-temperature excitation spectra of yttrium- (1) and gadolinium-based (2) hybrid materials synthesized in benzyl alcohol doped with Eu^{3+} and monitored around 625 nm. The dotted lines represent the room-temperature excitation spectra monitored around 611 nm for the same hybrid materials calcined at 800°C. **b** Room-temperature emission spectra of the yttrium- (1) and gadolinium-based (2) hybrid materials synthesized in benzyl alcohol doped with Eu^{3+} and excited under 284 nm

to the excitation of the phenyl rings of the benzoate molecules in between the layers and of a shoulder at ca. 260 nm attributed to the charge transfer band (CTB) related to the excitation of an electron from the 2p orbital of the oxygen to the $4f^6$ orbital. The presence of the phenyl rings excited states demonstrates that also in this case the benzoate complex plays an important role for the emission characteristics. The CTB in the excitation spectra of both materials can be selectively observed by monitoring the Eu^{3+} excitation spectra of the same samples calcined at 800°C (dotted lines in Figure 8.5a). The heat treatment removes the organic phase leading to a purely inorganic RE sesquioxide. The emission spectra excited for the wavelength that maximizes the Eu^{3+} emission intensity (284 nm, excitation into the phenyl rings of the benzoate molecules) are shown in Figure 8.5b. The emission spectra display a series of lines ascribed to the Eu^{3+} intra-$4f^6$ $^5D_0 \rightarrow {}^7F_{0-4}$ electronic transitions and no emission from the excited states of the phenyl rings could be detected (not shown), indicating an efficient energy transfer between them and the Eu^{3+} levels.

The emission of the hybrid materials doped with Eu^{3+} ions was further quantified through the measurement of the radiance and quantum yield under the excitation wavelength that maximizes the Eu^{3+} emission intensity. The radiance values found for the Eu^{3+}-doped hybrid materials are larger than the values of red standard phosphors (ca. 8–23% larger), even though the quantum yield of the Eu^{3+}-doped samples is lower than the one of the standard phosphors (ca. 15% compared to 95%). The larger radiance of the Eu^{3+}-doped hybrid materials can be explained by the larger absorption cross section of the phenyl rings compared to the CTB observed in purely inorganic samples.

As described in Chapter 4.2, hybrid materials could also be prepared using biphenylmethanol as solvent instead of benzyl alcohol. Their optical properties are in general similar to the ones presented above. The main difference is due to the larger absorption band characteristic of the biphenolate molecules that permits to observe RE emission by exciting the hybrid materials at a lower energy compared to the same materials synthesized in benzyl alcohol [43, 73].

Another interesting feature of these hybrid materials is due to their emission chromaticity that can be simply controlled by varying the excitation wavelength [86]. For example, the emission color of the Eu^{3+}-doped samples may be tuned from the red spectral region to the white one crossing the purplish-pink, bluish-purple and the purple spectral areas of the diagram when the excitation wavelength varies from 280 to 420 nm (Figure 8.6). It was thus possible to precisely tune the emission color through the CIE chromaticity diagram by varying the excitation wavelength and without loosing the high radiance values.

In conclusion organic-inorganic hybrid materials synthesized by the "benzyl alcohol route" show enhanced and new interesting emission properties compared to pure inorganic RE_2O_3 nanoparticles synthesized via other non-

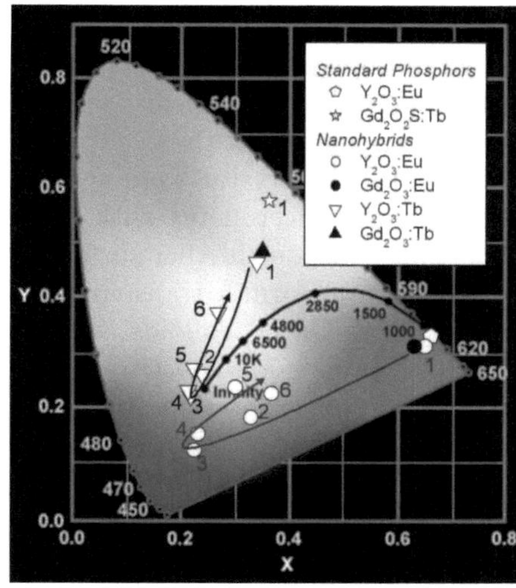

Fig. 8.6 CIE chromaticity diagram (1931) showing the (x,y) color coordinates of the room-temperature emission of yttrium- and gadolinium-based hybrid materials synthesized in benzyl alcohol doped with Tb^{3+} and Eu^{3+} ions and excited at different excitation wavelengths, 264-280 nm (1), 330 nm (2), 350 nm (3), 375 nm (4), 395 nm (5), and 420 nm (6) [86]. The black curve represents the color of the black body emission at different T (K)

aqueous routes. In fact, they not only show larger quantum yields (about one order of magnitude larger) but, due to the larger absorption cross section of the organic moieties, they show radiances nearly two orders of magnitude larger hence, also larger than commercial phosphors. However, the hybrid materials present a major disadvantage that prevent their use in biological analysis (e.g. labeling), i.e, their particle size could not be controlled as in the case of pure inorganic nanoparticles and it still remains a challenge.

8.3.2 Semiconductor Nanoparticles

Fluorescent semiconductor nanocrystals or quantum dots (especially II–VI and III–V) have attracted intensive interest during the past decade because their outstanding optical properties make them suitable for various fields such as biological analysis, lighting and photovoltaic applications. Indeed, they present strong size-dependent optical absorption and emission, the so called "quantum size effect" and high QY. Metal oxide nanoparticles such as zinc oxide also present interesting QY in view of such potential applications. However, they were synthesized using approaches not covered by the scope of this book [106, 107, 55]. The photoluminescence of pure semiconducting metal oxides generally arises from the recombination of charges trapped by defects (e.g. oxygen vacancies) which create energy levels inside the band gap and not from a direct recombination of the charge carriers as in II–VI and III–V direct band gap semiconductors. Therefore, even though the emission wavelength can be generally modified by the particle size, its wavelength is substantially larger than the one required to promote an electron from the valence band to the conduction band. Moreover, due to the physical origin (charge recombination from trap states) the width of the emission bands are much larger for an equivalent particle size distribution, than in the case of a direct recombination of the exciton.

$ZnGa_2O_4$ nanocrystal were recently prepared by solvothermally reacting metal acetylacetonates with benzylamine at 200°C [7]. The resulting nanocrystals without further surface modification could be readily dispersed in chloroform to form a transparent colloidal solution. The excitation spectrum of the $ZnGa_2O_4$ colloids is characterized by a broad peak centered at 330 nm. The emission spectrum shows a broad emission band centered at 427 nm under an excitation at 330 nm. This band is blue shifted compared to previous studies on $ZnGa_2O_4$ which reported the maximum of emission around 450–470 nm. This blue shift was attributed to a quantum size effect arising from the nanometric particle size. The QY was estimated to be 6.8% under an excitation at 347 nm. It is likely that the blue emission of $ZnGa_2O_4$ originates from the self-activation center of the octahedral Ga-O environment in the spinel structure of $ZnGa_2O_4$ from which the Ga^{3+} ions combine with UV-generated free electrons produced in oxygen vacancies [7].

8.4 (Photo)catalysis

The good accessibility of the surface of metal oxide nanoparticles prepared via surfactant-free routes should be particularly advantageous for applications in (photo)catalysis and gas sensor technology. Several studies proved that surfactants adsorbed on the surface of titania strongly influenced its photocatalytic activity [112, 95, 13]. Especially surfactants (e.g. sodium dodecylsulfate) that compete with the pollutant for degradation on the surface of titania lowered the photocatalytic activity considerably [112]. Accordingly, metal oxide nanoparticles synthesized in surfactant-free media showed excellent photocatalytical properties. Titania nanoparticles prepared from $TiCl_4$ and benzyl alcohol exhibited higher activity than the standard Degussa P25 photocatalyst for phenol degradation [118]. Although thermal treatment at 400°C increased the photocatalytic activity to a maximum due to the removal of organic residues on the nanoparticle surface, the temperature is still low enough to keep the nanocrystal size small. In comparison, trioctylphosphine oxide (TOPO), a frequently used surfactant, starts to decompose at around 425°C [29].

Comparelli et al. [18] compared the photocatalytic degradation of two azo dyes by two kind of organic-capped TiO_2 anatase nanoparticles and to the commercial Degussa P25 photocatalyst. The titania nanocrystals were either synthesized using TOPO as ligand in nonaqueous media following the route published by Trentler et al. [101] or using oleic acid as ligand by hydrolysis of $Ti(O\textit{i}Pr)_4$ by *in situ* generated water as it was reported previously by Cozzoli et al. [19]. On the one hand, they attribute the slightly higher activity of the oleic acid-capped nanocrystals, compared to the commercial P25 photocatalyst, to their smaller size and consequently large specific surface area. On the other hand, they found that TOPO-capped nanocrystals were the less effective in the photodegradation processes studied than the commercial P25 photocatalyst and the oleic acid-capped nanocrystals. This behavior was attributed to the role of the TOPO that almost fully passivates the surface. The strong Ti-O-P bonds are difficult to hydrolyze [18], underlining the detrimental role of surfactants and strong binding agents in applications such as catalysis or gas sensing requiring accessible surfaces.

Particularly high photocatalytic activity in the mineralization of acetic acid in aqueous solutions was achieved over titania nanoparticles synthesized from alkoxides in different alcohols without any surfactant [47, 48].

During the intense search for photocatalysts active under visible light, $InNbO_4$ nanoparticles of about 30 nm were recently prepared in benzyl alcohol [114]. Their activity in the degradation of rhodamine B in aqueous solution under visible light was found to be higher than the standard photocatalyst P25 and bulk $InNbO_4$ synthesized by traditional high temperature route.

In heterogeneous catalysis V_2O_3 made from vanadium isopropoxide in benzyl alcohol [74] is a good catalysts for the oxidation of n-butane towards maleic anhydride [35]. The peculiarity of this catalyst lies in its crystal structure, because the nanoparticle core consists of catalytically inactive V_2O_3, whereas the surface is rather complex with vanadium species of higher oxidation states (IV and V) that are known to be responsible for the catalytic reactivity of vanadium based catalysts in partial oxidation reactions.

Iron oxide nanoparticles synthesized in benzyl alcohol [77] were shown to effectively catalyze the hydrothermal carbonization of starch and rice grains under mild conditions (200°C) [20].

Cu_2O coated Cu nanoparticles prepared from copper acetylacetonate in oleylamine were used as catalyst for Ullmann type amination coupling reactions of aryl chlorides [96].

8.5 Gas Sensing

8.5.1 Introduction

Resistive chemical sensors based on metal oxide semiconductors (MOS) are attracting significant attention due to their simplicity, low cost, small size and ability to be integrated in electronic devices. Applications of MOS sensors can be found in many areas, such as in industry, environment, home safety, biomedicine, automotive and security. However, continuous improvements in the sensors properties are required for their implementation in constantly more demanding applications. Especially, better sensitivity and selectivity, faster response, together with low power consumption and high device reliability are sought after [109, 91, 110].

Resistive chemical sensors are devices in which the electrical transport properties of the MOS sensing layer are modified upon contact with reducing or oxidizing gases. The performance of a sensor is closely related to the nature and the structure of the sensing material and to the nanoscale morphology. The demand for high performance gas sensors generated strong incentives for the promotion of basic studies in different fields of materials sciences (synthesis, surface science, catalysis, etc.). In particular, many interesting results have been achieved and intensive efforts are still in progress in order to develop new materials and synthesis approaches [89, 98, 103, 28].

The gas sensing efficiency of MOS changes drastically between the bulk and nanostructured form. In fact, the same material having different size, shape and exposed facets (e.g. coarse grains, nanoparticles, nanowires, etc.) often shows dramatically different electric [5, 10, 49, 58], sensing [111, 3, 108, 53] and catalytic behavior [88, 31]. The correlation between these parameters and the sensing characteristic of MOS are, however, still not completely defined.

Nanocrystalline metal oxides are suitable for applications in catalysis and gas sensors as their properties are enhanced because of the much larger surface-to-bulk ratio compared to coarse micrograined materials. The present status of new nanostructured materials and their applications to resistive gas sensors has been recently reviewed [6, 9, 94]. In this section, we report the gas sensing properties of various semi-conducting metal oxide nanoparticles synthesized in organic solvents [67].

8.5.2 Sensor Devices

Gas sensor devices having sensing layers made of metal oxide nanoparticles are generally fabricated by simple drop/spin coating or screen-printing of a nanoparticle suspension onto a patterned substrate equipped with electric contacts on the front and a heater on the back. Schematic representation of such a device is given in Figure 8.7. The deposited layer is then thermally treated at moderate temperature in order to eliminate the binder, and to form a porous structure attached to the substrate. The principles of sensing mechanism of resistive gas sensors are reported in many books and reviews and will not be considered here in details [34, 60, 1, 4, 87, 104]. In brief, the response of a n-type metal oxide semiconductor to the presence of a target gas relies on the surface reactions which occur between adsorbed oxygen species and the probed gas. Oxygen adsorbed on the surface traps free electrons because of its high electron affinity, forming a potential barrier at the grain boundaries. This potential barrier restricts the flow of electrons, causing the electric

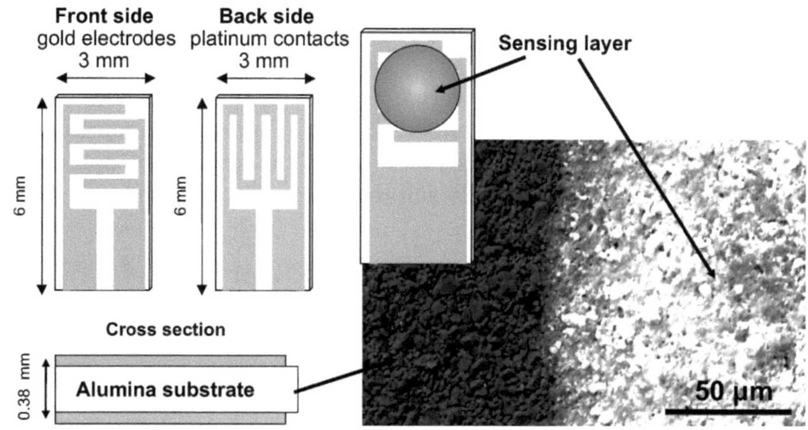

Fig. 8.7. Schematic representation of the MOS sensor device and micrograph of the In_2O_3 metal oxide sensing layer deposited on the ceramic substrate

resistance to increase. When the sensor is exposed to an atmosphere containing reducing gases, e.g. hydrocarbons, CO, ethanol, etc., the gas molecules adsorb on the surface and react with active oxygen species. Reactions with surface oxygen species vary depending on the temperature and the reactivity of the sensing material. In any case, this lowers the potential barrier allowing electrons to flow more easily, thereby reducing the electrical resistance. On the other hand, the adsorption of oxidizing gases such as NO_2 and ozone on metal oxide surfaces increases the surface resistance [50, 84]. The opposite is true for p-type oxides, where electron exchange due to the gas interaction leads either to a reduction (reducing gas) or an increase (oxidizing gas) in electron holes in the valence band [54, 85].

As the various pathways and reactions involved on the surface of the metal oxide semiconductors are complex and not always known, it is not simple to establish correlations between the electrical/gas sensing properties and the morphological and microstructural characteristics. Indeed, it should be considered that gas sensing responses depend on a number of parameters such as chemical (catalytic and acid-base properties of the surface), microstructural (particle size, crystal structure) and electronic (electronic structure, Fermi level position, transport properties) properties.

8.5.3 Nanoparticles Made in Surfactant-Free Systems

The response of various metal oxide nanoparticles, synthesized in surfactant free systems and especially in benzyl alcohol, to various oxidizing and reducing gases where recently reported [62, 63, 64, 65, 66, 75, 78, 80].

8.5.3.1 Tin-Indium Metal Oxides

The indium tin oxide system is very rich and gave rise to homogeneous nanocrystalline particles in the range 2-25 nm (Figure 8.8) depending on the composition and on the precursors used (cf. Chapter 4.1.4) [78, 2, 64, 65]. In_2O_3-based sensors are particularly suitable for monitoring nitrogen dioxide

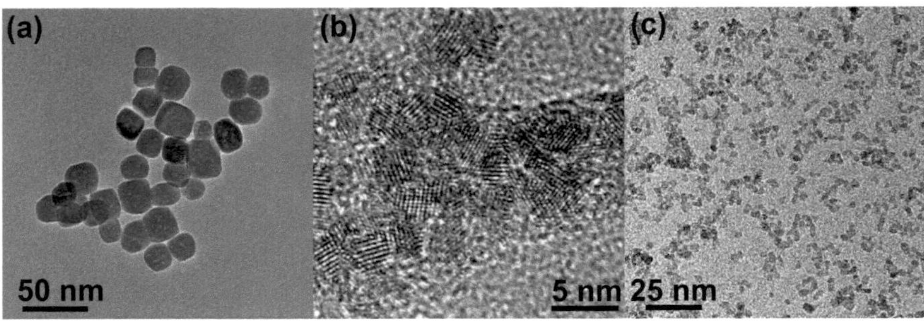

Fig. 8.8. **a** TEM image of an assembly of In_2O_3 cube-shaped nanoparticles, **b** SnO_2 and **c** $Sn_{0.9}In_{0.1}O_x$ nanoparticles

at ppb-ppm levels [38]. NO_2 is a highly reactive species showing a strong oxidizing character when adsorbed onto non-stoichiometric metal oxides. In fact, as metal oxide surfaces are expected to deviate from exact stoichiometry, an increase of surface sites for chemisorption of oxygen and oxidizing species are generally expected [30]. For this reason, non-stoichiometric oxides generally permit the detection of oxidizing gases at lower temperature compared to reducing gases.

The response of the In_2O_3 nanoparticles-based sensor to trace levels of NO_2 in function of the operating temperature is reported in Figure 8.9. The response to NO_2 increases with increasing temperature, showing a maximum of sensitivity $S = I_{air}/I_{2ppmNO_2} = 100$ around 200°C (Figure 8.9a). The dynamic response of the sensor at 250°C shows that the output signals essentially recover to the initial level, indicating the reversibility of the interactions between the sensing elements and NO_2, with fast (less than 60 sec.) response time (Figure 8.9b). The estimated detection limit of this sensor is about 1 ppb of NO_2 in dry air [78].

Gas sensors based on SnO_2 nanoparticles synthesized in benzyl alcohol showed higher responses to reducing gases (e.g. CO and CH_4) compared to In_2O_3 [65]. The gas sensing behavior of mixed indium tin oxide nanoparticles was also studied [65, 64] showing that the sensitivity and selectivity towards reducing and oxidizing gases depends on the chemical composition and on the working temperature of the sensor.

In_2O_3 nanoparticles were further doped with 1 wt% of platinum. The dopant was dispersed on the surface of the semiconducting metal oxide by

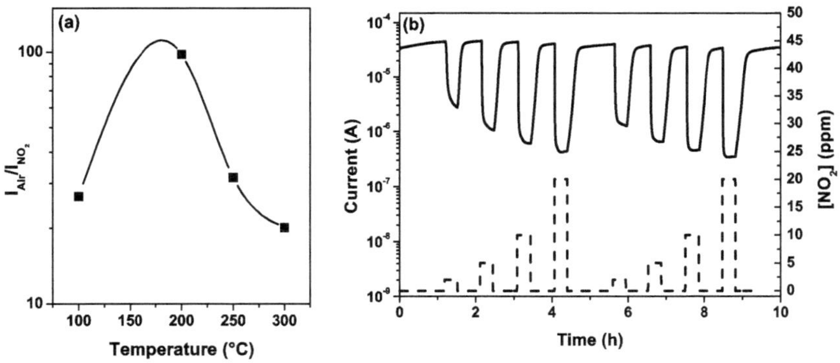

Fig. 8.9. a Gas responses as a function of working temperature to 2 ppm of NO_2, **b** dynamic response (*solid line*) of the In_2O_3-based sensor to different NO_2 concentrations (*dashed line*) in dry air

wetness impregnation, contacting the In_2O_3 nanopowders with the proper amount of an aqueous solution of H_2PtCl_6. The nanopowders were finally dried and fired at controlled temperature (between 150 and 450°C) to promote the decomposition of the platinum precursor into metallic platinum. The obtained material (Figure 8.10) is composed of cubic-shaped indium oxide nanoparticles with small platinum particles of about 2-3 nm attached to their surface. Sensor devices made of these platinum-doped In_2O_3 nanoparticles showed a high sensitivity to oxygen even at room temperature [62]. However, long response and recovery times were observed at low temperature. To be of any practical use the sensor should work at higher temperature (150°C). Figure 8.11 depicts the sensitivity of the Pt-promoted and the reference sensors towards 20% oxygen. It proves that platinum substantially increases the sensitivity. Moreover, faster response and recovery times are also observed [63].

8.5.3.2 Tungsten Oxides

Tungsten oxides are relevant semiconducting oxides for gas sensing applications and were also successfully synthesized in benzyl alcohol from tungsten chloride or alkoxide as precursors (cf. Chapter 4.2.3) [68, 79, 80, 81]. In function of the precursor and other ligands added during the synthesis it was possible to synthesize nanoplatelets, nanowires and hybrid materials. Especially, the tungsten oxide nanowires with a uniform diameter of about 1.3 nm and an aspect ratio of more than 500 (cf. Figure 4.12) showed to be promising for sensing applications [80, 66]. Tungsten oxide nanowires synthesized from WCl_6 in presence of deferoxamine mesylate (DFOM) showed high sensitivity towards ammonia (Figure 8.12a). The authors proposed to use it for

Fig. 8.10 TEM image of In_2O_3 nanoparticles doped with 3 nm Pt particles

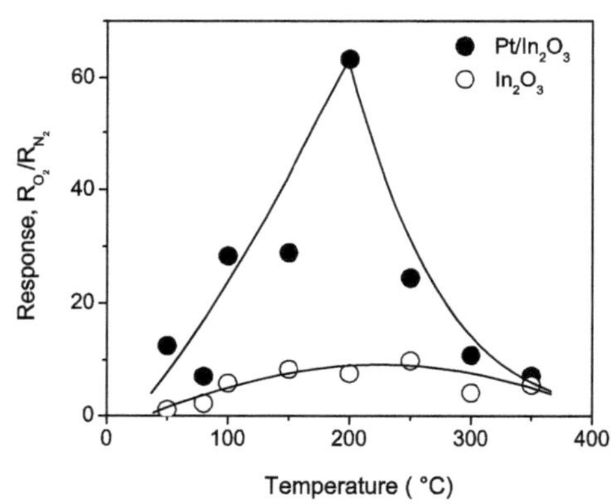

Fig. 8.11 Responses of the doped and undoped In_2O_3-based sensors to 20% of oxygen at different working temperatures

clinical breath analysis for the detection of ammonia [66]. Indeed, the sensor sensitivity to some interfering gases such as acetone, CO and NO in concentrations much higher than their usual concentration in the human breath was negligible, proving its possible application for healthcare applications (Figure 8.12b).

Similar tungsten oxide nanowire bundles synthesized from tungsten isopropoxide exhibited a particularly high sensitivity at low NO_2 concentrations. A sensor signal of S~9 was measured at low temperature for a concentration of only 50 ppb of NO_2 [80].

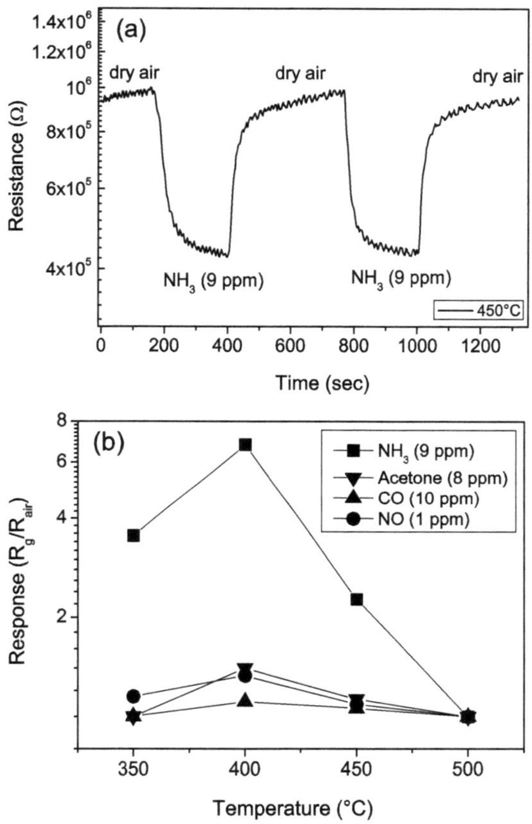

Fig. 8.12. a Transient response of the tungsten oxide nanowires-based sensor to 9 ppm of ammonia in dry air at 450°C. **b** Response of the tungsten oxide nanowires-based sensor to 9 ppm of NH_3 as a function of the operating temperature. The response to other interfering gases is also reported for comparison

8.5.3.3 $CdIn_2O_4$

$CdIn_2O_4$ nanoparticles with crystallite sizes of about 10 nm were obtained by reacting cadmium acetate and indium isopropoxide in benzyl alcohol without the use of any additional stabilizing agents for controlling the crystal growth [8]. The as-fabricated sensor showed high response as well as fast response and recovery times when exposed to ethanol vapors. At an operating temperature of 260°C, the response and recovery times were 6 and 30 s, respectively, for an ethanol concentration of 1000 ppm. The authors pointed out that their sensor shows better performances than previously reported ethanol sensors based on

SnO_2 and V_2O_5 nanostructures [56, 12, 57, 99]. Furthermore, cross selectivity tests were also performed and low responses to gasoline and butane was found. This result indicates that $CdIn_2O_4$ is a promising new gas-sensing material for detecting ethanol and has the prerequisites to be used for monitoring alcohols in various applications.

8.5.4 Nanoparticles Made in Surfactant Systems

Other synthesis approaches, even though not totally nonaqueous, were recently developed for the synthesis of metal oxide nanoparticles for gas sensing applications. For example, Epifani et al. [27] (cf. Chapter 3.1) synthesized nanoparticles using a two-step process involving a hydrolytic preparation of metal oxide sols and their subsequent low-temperature injection into a mixture of organic solvents and long chain amines as surfactants. The as fabricated nanoparticles were then drop coated on sensor devices and the gas sensing properties measured [27, 26]. SnO_2 and In_2O_3 just few nanometers in size possess good sensitivities to NO_2 at sub-ppm concentrations even at room temperature [27]. The same In_2O_3 nanoparticles with sizes from 4 to 10 nm exhibited enormous response to ozone gas, with a response of several orders of magnitude, even for concentrations as low as 60 ppb [26].

8.5.5 Sensing Layers Synthesized by ALD

More recently, nonaqueous sol-gel routes applied to ALD (cf. Chapter 4.3) were also used to deposit V_2O_4 layers onto carbon nanotubes (CNTs) [105]. CNTs coated with 4.5 nm of vanadium oxide are shown in Figure 8.13. The scanning electron microscope (SEM) image recorded using back scattered electrons (Figure 8.13a) shows brighter contours in the inside and outside of the tube. These regions of enhanced intensity are due to the stronger scattering of the heavier element (i.e. vanadium) thus, proving that the tubes are indeed coated with a material of larger electronic density. In the TEM image (Figure 8.13b) the darker regions on the outer and inner walls of the CNTs correspond to the metal oxide layers deposited by the process. The coating is uniform along the whole surface of the nanotubes and presents approximately the same thickness on the inner and outer surface.

The hybrid material was drop coated onto gas sensor devices and its response towards different gases was probed. It was observed that the exposition of uncoated CNTs to 6.5 ppm of NO_2 causes a decrease of the resistance which can be reversed upon exposure to dry air (Figure 8.14a). Thus, the pure CNTs show a typical p-type semiconducting behavior. On the other hand, V_2O_4-coated CNTs show an opposite sensing response. Indeed, their resistance increases upon exposure to NO_2, (Figure 8.14b) clearly indicating that V_2O_4-coated CNTs show n-type semiconducting character. The resis-

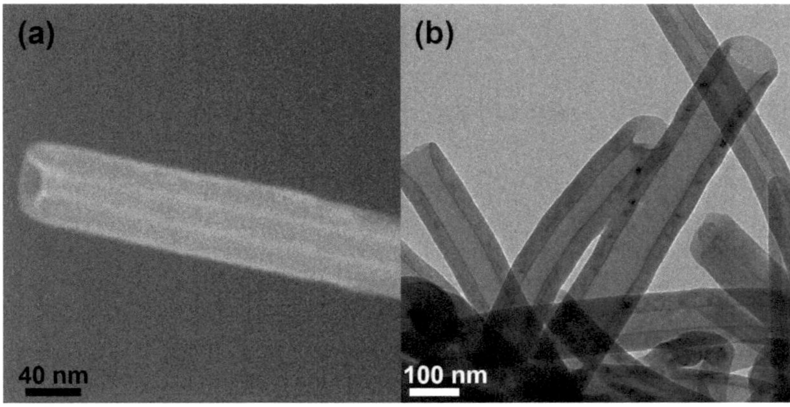

Fig. 8.13. **a** SEM and **b** TEM images of the CNTs coated by a 4.5 nm thick layer of vanadium oxide

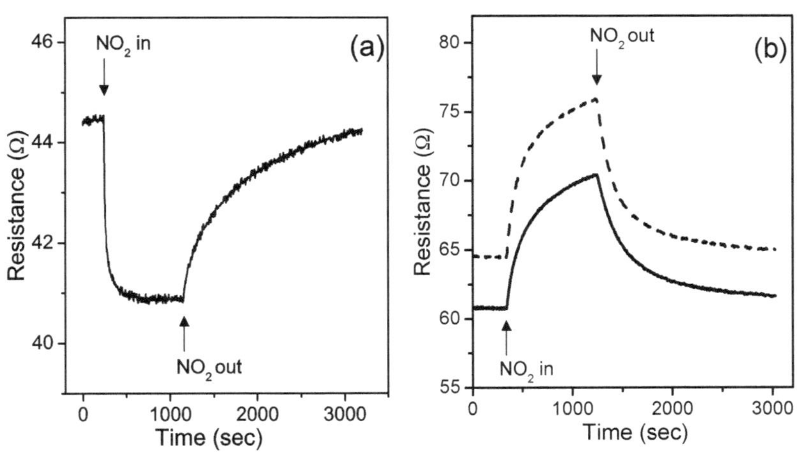

Fig. 8.14. Transient response of sensors pre-treated in air at 150°C. a) Uncoated CNTs, b) CNTs coated by 4.5 nm (*solid line*) and 2 nm (*dotted line*) thick layer of vanadium oxide

tance of sensors based on V_2O_4-CNTs remains very low, regardless of the presence of a V_2O_4 layer (Figure 8.14), suggesting that the current mainly flows through the p-type nanotube structure. Therefore, it was concluded that the heterostructure formed by the combination of the n-type film on top of the p-type CNT (n-V_2O_4/p-CNTs) is responsible for the enhanced and inverted response to NO_2 [105].

8.6 Biomedical Applications

Nanometer-sized nanoparticles are promising for diverse biomedical applications as a result of their physical (especially optical and magnetic) properties. As a typical example, semiconductor nanoparticles are currently used as fluorescence probes for labeling and optical imaging of biological tissues (cf. for example Ref. [59]). Magnetic nanoparticles find their application in magnetic resonance imaging (MRI) as contrast-enhancement agents. They can bring advantages over currently used MRI contrast agents (e.g. gadolinium complexes) such as better contrast and selectivity to a particular tissue or cell type. Several reports making use of magnetic metal oxide nanoparticles, synthesized in organic solvents, for MRI application have been published and were recently reviewed [42].

MnO nanoparticles synthesized from the thermal decomposition of Mn-oleate complex were subsequently made biocompatible by encapsulating them in a polyethylene glycol (PEG)-phospholipid shell [61] in a similar way as it was previously achieved for quantum dots [25]. They were highly crystalline, stable in water and showed no degradation or agglomeration over several months. It was shown that the nanoparticle size plays an important role on the MRI contrast. The brightest signal in the T_1-weighted MR images[1] was observed for the smallest particles studied (7 nm). It was reported that the MnO nanoparticles clearly decrease the longitudinal and the transverse relaxation times (T_1 and T_2, respectively) showing to be potentially useful as contrast agents in MRI.

Magnetite nanoparticles synthesized from a similar approach [71] were incorporated into poly(D,L-lactic-co-glycolic acid) (PLGA) nanoparticles and were used for magnetically guided delivery and T_2 MRI contrast agent [46].

12 nm magnetite nanocrystals synthesized from $Fe(CO)_5$ [11] were also used as MRI contrast agents after functionalization with hydrophilic ligands such as (3-carboxypropyl)trimethylammonium chloride and 2-carboxyethyl phosphonate permitting their dispersion in aqueous media. They showed to be noncytotoxic, and highly efficient magnetic nanocrystal probes for intracellular labeling and in vivo MR tracking. By simple modulation of the nanocrystal surface charge properties (i.e. cationic or anionic ligands), it was possible to label a variety of cell types [97].
Magnetite and ferrite nanocrystals synthesized from $Fe(acac)_3$, as described

[1] T_1 is the longitudinal relaxation time

previously [100], were functionalized with 2,3-dimercaptosuccinic acid (DMSA) and conjugated with the cancer-targeting antibody (Herceptin) for in-vivo and in-vitro breast cancer diagnosis by MRI [41, 36, 51]. These DMSA coated nanoparticles were also used to fabricate multifunctional probes comprised of a dye-doped silica core and multiple satellites of magnetic nanoparticles. Their utilization as simultaneous optical and MRI of neuroblastoma cells expressing polysialic acids was demonstrated [52]. The same particles were also used to hybridize adenoviruses in order to probe their viral gene delivery [37].

Finally, TiO_2 nanoparticles synthesized in a mixture of oleic acid, oleylamine and stabilized in water after functionalization with DMSA were reported to be photocatalytically active in skin cancer cell treatment [90].

8.7 Other Applications

Other examples of applications of metal oxide nanoparticles synthesized in organic solvents were also presented in the recent literature.

ZnO nanoparticles synthesized in diethylene glycol from zinc acetate dihydrate following the method of Jezequel et al. [40, 39] were used in dye-sensitized solar cells [115, 116, 14]. In this system the primary nanoparticles aggregate in spherical superstructures promoting light scattering. This leads to the enhancement of the photon absorption and to the improvement of the short-circuit current density and of the overall light conversion efficiencies.

ZrO_2 nanoparticles synthesized in benzyl alcohol were used for the fabrication of organic-inorganic nanocomposites that can be selectively photopolymerized to inscribe extremely effective volume holographic gratings with the highest refractive index contrast (n_1 of up to 0.024) achieved so far [33].

References

1. Azad, A.M., Akbar, S.A., Mhaisalkar, S.G., Birkefeld, L.D., Goto, K.S.: Solid-state gas sensors: A review. J. Electrochem. Soc. **139**, 3690–3704 (1992)
2. Ba, J.H., Fattakhova-Rohlfing, D., Feldhoff, A., Brezesinski, T., Djerdj, I., Wark, M., Niederberger, M.: Nonaqueous synthesis of uniform indium tin oxide nanocrystals and their electrical conductivity in dependence of the tin oxide concentration. Chem. Mater. **18**, 2848–2854 (2006)
3. Baraton, M.I., Merhari, L.: Influence of the particle size on the surface reactivity and gas sensing properties of SnO_2 nanopowders. Mater. Trans. **42**, 1616–1622 (2001)
4. Barsan, N., Schweizer-Berberich, M., Gopel, W.: Fundamental and practical aspects in the design of nanoscale SnO_2 gas sensors: A status report. Fresenius J. Anal. Chem. **365**, 287–304 (1999)
5. Bose, A.C., Thangadurai, P., Ramasamy, S.: Grain size dependent electrical studies on nanocrystalline SnO_2. Mater. Chem. Phys. **95**, 72–78 (2006)
6. Cao, G.: Nanostructures and nanomaterials. IC Press, London (2004)

7. Cao, M., Djerdj, I., Antonietti, M., Niederberger, M.: Non-aqueous synthesis of colloidal $ZnGa_2O_4$ nanocrystals and their photoluminescence properties. Chem. Mater. **19**, 5830–5832 (2007)
8. Cao, M., Wang, Y., Chen, T., Antonietti, M., Niederberger, M.: A highly sensitive and fast-responding ethanol sensor based on $CdIn_2O_4$ nanocrystals synthesized by a nonaqueous sol-gel route. Chem. Mater. **20**, 5781–5786 (2008)
9. Capone, S., Siciliano, P.: Gas sensors from nanostructured metal oxides, *Encyclopedia of Nanoscience and Nanotechnology*, vol. 3. American Scientific Publishers (2004)
10. Carcia, P.F., Ferretti, A., Suna, A.: Particle size effects in thick film resistors. J. Appl. Phys. **53**, 5282–5288 (1982)
11. Cheon, J., Kang, N.J., Lee, S.M., Lee, J.H., Yoon, J.H., Oh, S.J.: Shape evolution of single-crystalline iron oxide nanocrystals. J. Am. Chem. Soc. **126**, 1950–1951 (2004)
12. Chiu, H.C., Yeh, C.S.: Hydrothermal synthesis of SnO_2 nanoparticles and their gas-sensing of alcohol. J. Phys. Chem. C **111**, 7256–7259 (2007)
13. Choi, H., Stathatos, E., Dionysiou, D.D.: Effect of surfactant in a modified sol on the physicochemical properties and photocatalytic activity of crystalline TiO_2 nanoparticles. Top. Catal. **44**, 513–521 (2007)
14. Chou, T.P., Zhang, Q., Fryxell, G.E., Cao, G.Z.: Hierarchically structured ZnO film for dye-sensitized solar cells with enhanced energy conversion efficiency. Adv. Mater. **19**, 2588–2592 (2007)
15. Clavel, G., Pinna, N., Zitoun, D.: Magnetic properties of cobalt and manganese doped ZnO nanowires. Phys. Stat. Sol. A **204**, 118–124 (2007)
16. Clavel, G., Willinger, M.G., Zitoun, D., Pinna, N.: Solvent dependent shape and magnetic properties of doped ZnO nanostructures. Adv. Funct. Mater. **17**, 3159–3169 (2007)
17. Clavel, G., Willinger, M.G., Zitoun, D., Pinna, N.: Manganese-doped zirconia nanocrystals. Eur. J. Inorg. Chem. pp. 863–868 (2008)
18. Comparelli, R., Fanizza, E., Curri, M.L., Cozzoli, P.D., Mascolo, G., Passino, R., Agostiano, A.: Photocatalytic degradation of azo dyes by organic-capped anatase TiO_2 nanocrystals immobilized onto substrates. Appl. Catal., B **55**, 81–91 (2005)
19. Cozzoli, P.D., Kornowski, A., Weller, H.: Low-temperature synthesis of soluble and processable organic-capped anatase TiO_2 nanorods. J. Am. Chem. Soc. **125**, 14,539–14,548 (2003)
20. Cui, X., Antonietti, M., Yu, S.H.: Structural effects of iron oxide nanoparticles and iron ions on the hydrothermal carbonization of starch and rice carbohydrates. Small **2**, 756–759 (2006)
21. Dietl, T., Ohno, H., Matsukura, F., Cibert, J., Ferrand, D.: Zener model description of ferromagnetism in zinc-blende magnetic semiconductors. Science **287**, 1019–1022 (2000)
22. Djerdj, I., Arcon, D., Jaglicic, Z., Niederberger, M.: Nonaqueous synthesis of metal oxide nanoparticles: Short review and doped titanium dioxide as case study for the preparation of transition-metal doped oxide nanoparticles. J. Solid State Chem. **181**, 1574–1584 (2008)
23. Djerdj, I., Garnweitner, G., Arcon, D., M., P., Zvonko, J., Niederberger, M.: Diluted magnetic semiconductors: Mn/Co-doped ZnO nanorods as case study. J. Mater. Chem. **18**, 5208–5217 (2008)
24. Dormann, J., Fiorani, D., Tronc, E.: Advances in chemical physics, chap. Magnetic relaxation in fine-particle systems, pp. 283–494. John Wiley & Sons Inc. (1997)
25. Dubertret, B., Skourides, P., Norris, D.J., Noireaux, V., Brivanlou, A.H., Libchaber, A.: In vivo imaging of quantum dots encapsulated in phospholipid micelles. Science **298**, 1759–1762 (2002)

26. Epifani, M., Comini, E., Arbiol, J., Pellicer, E., Siciliano, P., Faglia, G., Morante, J.R.: Nanocrystals as very active interfaces: Ultrasensitive room-temperature ozone sensors with In_2O_3 nanocrystals prepared by a low-temperature sol-gel process in a coordinating environment. J. Phys. Chem. C **111**, 13,967–13,971 (2007)
27. Epifani, M., Díaz, R., Arbiol, J., Comini, E., Sergent, N., Pagnier, T., Siciliano, P., Faglia, G., Morante, J.R.: Nanocrystalline metal oxides from the injection of metal oxide sols in coordinating solutions: Synthesis, characterization, thermal stabilization, device processing, and gas-sensing properties. Adv. Funct. Mater. **16**, 1488–1498 (2006)
28. Epifani, M., Diaz, R., Arbiol, J., Siciliano, P., Morante, J.R.: Solution synthesis of thin films in the SnO_2-In_2O_3 system: A case study of the mixing of sol-gel and metal-organic solution processes. Chem. Mater. **18**, 840–846 (2006)
29. Foos, E.E., Wilkinson, J., Makinen, A.J., Watkins, N.J., Kafafi, Z.H., Long, J.P.: Synthesis and surface composition study of CdSe nanoclusters prepared using solvent systems containing primary, secondary, and tertiary amines. Chem. Mater. **18**, 2886–2894 (2006)
30. Forleo, A., Francioso, L., Epifani, M., Capone, S., Taurino, A.M., Siciliano, P.: NO_2-gas-sensing properties of mixed In_2O_3-SnO_2 thin films. Thin Solid Films **490**, 68–73 (2005)
31. Freund, H.J., Libuda, J., Baumer, M., Risse, T., Carlsson, A.: Cluster, facets, and edges: Site-dependent selective chemistry on model catalysts. Chem. Record **3**, 181–200 (2003)
32. Furdyna, J.K.: Diluted magnetic semiconductors. J. Appl. Phys. **64**, R29–R64 (1988)
33. Garnweitner, G., Goldenberg, L.M., Sakhno, O.V., Antonietti, M., Niederberger, M., Stumpe, J.: Large-scale synthesis of organophilic zirconia nanoparticles and their application in organic-inorganic nanocomposites for efficient volume holography. Small **3**, 1626–1632 (2007)
34. Gurlo, A., Barsan, N., Weimar, U.: Gas sensors based on semiconductiong metal oxides. Metal Oxides: Chemistry and Applications. CRC Press, Boca Raton (2006)
35. Havecker, M., Pinna, N., Weiss, K., Sack-Kongehl, H., Jentoft, R.E., Wang, D., Swoboda, M., Wild, U., Niederberger, M., Urban, J., Su, D.S., Schlogl, R.: Synthesis and functional verification of the unsupported active phase of V_xO_y catalysts for partial oxidation of n-butane. J. Catal. **236**, 221–232 (2005)
36. Huh, Y.M., Jun, Y.w., Song, H.T., Kim, S., Choi, J.s., Lee, J.H., Yoon, S., Kim, K.S., Shin, J.S., Suh, J.S., Cheon, J.: In vivo magnetic resonance detection of cancer by using multifunctional magnetic nanocrystals. J. Am. Chem. Soc. **127**, 12,387–12,391 (2005)
37. Huh, Y.M., Lee, E.S., Lee, J.H., wook Jun, Y., Kim, P.H., Yun, C.O., Kim, J.H., Suh, J.S., Cheon, J.: Hybrid nanoparticles for magnetic resonance imaging of target-specific viral gene delivery. Adv. Mater. **19**, 3109–3112 (2007)
38. Ivanovskaya, M., Gurlo, A., Bogdanov, P.: Mechanism of O_3 and NO_2 detection and selectivity of In_2O_3 sensors. Sens. Actuators B **77**, 264–267 (2001)
39. Jezequel, D., Guenot, J., Jouini, N., Fievet, F.: Preparation and morphological characterization of fine, spherical, monodisperse particles of ZnO. Mater. Sci. Forum **152-153**, 339–342 (1994)
40. Jezequel, D., Guenot, J., Jouini, N., Fievet, F.: Submicrometer zinc oxide particles: Elaboration in polyol medium and morphological characteristics. J. Mater. Res. **10**, 77–83 (1995)
41. Jun, Y.w., Huh, Y.M., Choi, J.s., Lee, J.H., Song, H.T., Kim S., Yoon, S., Kim, K.S., Shin, J.S., Suh, J.S., Cheon, J.: Nanoscale size effect of magnetic nanocrystals and their utilization for cancer diagnosis via magnetic resonance imaging. J. Am. Chem. Soc. **127**, 5732–5733 (2005)

42. Jun, Y.W., Lee, J.S., Cheon, J.: Chemical design of nanoparticle probes for high-performance magnetic resonance imaging. Angew. Chem. Int. Ed. **47**, 5122–5135 (2008)
43. Karmaoui, M., Mafra, L., Sá Ferreira, R.A., Rocha, J., Carlos, L.D., Pinna, N.: Photoluminescent rare-earth based biphenolate lamellar nanostructures. J. Phys. Chem. C **111**, 2539–2544 (2007)
44. Karmaoui, M., Sá Ferreira, R.A., Carlos, L.D., Pinna, N.: Lanthanide-based lamellar nanohybrids: The case of erbium. Mater. Sci. Eng. C **27**, 1368–1371 (2007)
45. Karmaoui, M., Sá Ferreira, R.A., Mane, A.T., Carlos, L.D., Pinna, N.: Lanthanide-based lamellar nanohybrids: Synthesis, structural characterization, and optical properties. Chem. Mater. **18**, 4493–4499 (2006)
46. Kim, J., Lee, J.E., Lee, S.H., Yu, J.H., Lee, J.H., Park, T.G., Hyeon, T.: Designed fabrication of a multifunctional polymer nanomedical platform for simultaneous cancer-targeted imaging and magnetically guided drug delivery. Adv. Mater. **20**, 478–483 (2008)
47. Kominami, H., Kato, J., Murakami, S., Kera, Y., Inoue, M., Inui, T., Ohtani, B.: Synthesis of titanium IV oxide of ultra-high photocatalytic activity: High-temperature hydrolysis of titanium alkoxides with water liberated homogeneously from solvent alcohols. J. Mol. Catal. A: Chem. **144**, 165–171 (1999)
48. Kominami, H., Kato, J., Takada, Y., Doushi, Y., Ohtani, B., Nishimoto, S., Inoue, M., Inui, T., Kera, Y.: Novel synthesis of microcrystalline titanium(IV) oxide having high thermal stability and ultra-high photocatalytic activity: thermal decomposition of titanium(IV) alkoxide in organic solvents. Catal. Lett. **46**, 235–240 (1997)
49. Lavik, E.B., Kosacki, I., Tuller, H.L., Chiang, Y.M., Ying, J.Y.: Nonstoichiometry and electrical conductivity of nanocrystalline CeO_{2-x}. J. Electroceram. **1**, 7–14 (1997)
50. Leblanc, E., Perier-Camby, L., Thomas, G., Gibert, R., Primet, M., Gelin, P.: NO_x adsorption onto dehydroxylated or hydroxylated tin dioxide surface. Application to SnO_2-based sensors. Sens. Actuators B **62**, 67–72 (2000)
51. Lee, J.H., Huh, Y.M., wook Jun, Y., wook Seo, J., tak Jang, J., Song, H.T., Kim, S., Cho, E.J., Yoon, H.G., Suh, J.S., Cheon, J.: Artificially engineered magnetic nanoparticles for ultra-sensitive molecular imaging. Nature Med. **13**, 95–99 (2007)
52. Lee, J.H., wook Jun, Y., Yeon, S.I., Shin, J.S., Cheon, J.: Dual-mode nanoparticle probes for high-performance magnetic resonance and fluorescence imaging of neuroblastoma. Angew. Chem. Int. Ed. **45**, 8160–8162 (2006)
53. Li, G.J., Zhang, X.H., Kawi, S.: Relationships between sensitivity, catalytic activity, and surface areas of SnO_2 gas sensors. Sens. Actuators B **60**, 64–70 (1999)
54. Li, Y., Wlodarski, W., Galatsis, K., Moslih, S.H., Cole, J., Russo, S., Rockelmann, N.: Gas sensing properties of p-type semiconducting Cr-doped TiO_2 thin films. Sens. Actuators B **83**, 160–163 (2002)
55. Liu, D.P., Li, G.D., Su, Y., Chen, J.S.: Highly luminescent ZnO nanocrystals stabilized by ionic-liquid components. Angew. Chem., Int. Ed. **45**, 7370–7373 (2006)
56. Liu, J., Wang, X., Peng, Q., Li, Y.: Vanadium pentoxide nanobelts: Highly selective and stable ethanol sensor materials. Adv. Mater. **17**, 764–767 (2005)
57. Liu, Y., Koep, E., Liu, M.: A highly sensitive and fast-responding SnO_2 sensor fabricated by combustion chemical vapor deposition. Chem. Mater. **17**, 3997–4000 (2005)
58. McAleer, J.F., Moseley, P.T., Norris, J.O.W., Williams, D.E.: Tin dioxide gas sensors. Part 1. Aspects of the surface chemistry revealed by electrical conductance variations. J. Chem. Soc. Faraday Trans. **83**, 1323–1346 (1987)

59. Michalet, X., Pinaud, F.F., Bentolila, L.A., Tsay, J.M., Doose, S., Li, J.J., Sundaresan, G., Wu, A.M., Gambhir, S.S., Weiss, S.: Quantum dots for live cells, in vivo imaging, and diagnostics. Science **307**, 538–544 (2005)
60. Moseley, P.T., Norris, J. (eds.): Techniques and mechanisms in gas sensing. IOP Publication Ltd, Bristol (1991)
61. Na, H.B., Lee, J.H., An, K., Park, Y.I., Park, M., Lee, I.S., Nam, D.H., Kim, S.T., Kim, S.H., Kim, S.W., Lim, K.H., Kim, K.S., Kim, S.O., Hyeon, T.: Development of a T1 contrast agent for magnetic resonance imaging using MnO nanoparticles. Angew. Chem. Int. Ed. **46**, 5397–5401 (2007)
62. Neri, G., Bonavita, A., Micali, G., Rizzo, G., Galvagno, S., Niederberger, M., Pinna, N.: A highly sensitive oxygen sensor operating at room temperature based on platinum-doped In_2O_3 nanocrystals. Chem. Commun. pp. 6032–6034 (2005)
63. Neri, G., Bonavita, A., Micali, G., Rizzo, G., Pinna, N., Niederberger, M.: In_2O_3 and Pt-In_2O_3 nanopowders for low temperature oxygen sensors. Sens. Actuators B **127**, 445–462 (2007)
64. Neri, G., Bonavita, A., Micali G. Rizzo, G., Pinna, N., Niederberger, M., Ba, J.: Effect of the chemical composition on the sensing properties of In_2O_3-SnO_2 nanoparticles synthesized by a non-aqueous method. Sens. Actuators B **130**, 222–230 (2008)
65. Neri, G., Bonavita, A., Rizzo, G., Galvagno, S., Pinna, N., Niederberger, M., Capone, S., Siciliano, P.: Towards enhanced performances in gas sensing: SnO_2 based nanocrystalline oxides application. Sens. Actuators B **122**, 564–571 (2007)
66. Neri, G., Micali G. Bonavita, A., Ipsale, S., Rizzo, G., Niederberger, M., Pinna, N.: Tungsten oxide nanowires-based ammonia gas sensors. Sens. Lett. **6**, 590–595 (2008)
67. Neri, G., Pinna, N.: Nanomaterials: new research developments, chap. Resistive chemical sensors from metal oxides nanocrystals synthesized in organic solvents, pp. 217–252. Nova Science Publishers (2008)
68. Niederberger, M., Bartl, M.H., Stucky, G.D.: Benzyl alcohol and transition metal chlorides as a versatile reaction system for the nonaqueous and low-temperature synthesis of crystalline nano-objects with controlled dimensionality. J. Am. Chem. Soc. **124**, 13,642–13,643 (2002)
69. Ostanin, S., Ernst, A., Sandratskii, L.M., Bruno, P., Dane, M., Hughes, I.D., Staunton, J.B., Hergert, W., Mertig, I., Kudrnovsky, J.: Mn-stabilized zirconia: From imitation diamonds to a new potential high-Tc ferromagnetic spintronics material. Phys. Rev. Lett. **98**, 016,101 (2007)
70. Panda, A.B., Glaspell, G., El-Shall, M.S.: Microwave synthesis and optical properties of uniform nanorods and nanoplates of rare earth oxides. J. Phys. Chem. C **111**, 1861–1864 (2007)
71. Park, J., An, K., Hwang, Y., Park, J.G., Noh, H.J., Kim, J.Y., Park, J.H., Hwang, N.M., Hyeon, T.: Ultra-large-scale syntheses of monodisperse nanocrystals. Nature Mater. **3**, 891–895 (2004)
72. Park, J., Lee, E., Hwang, N.M., Kang, M., Kim, S.C., Hwang, Y., Park, J.G., Noh, H.J., Kim, J.Y., Park, J.H., Hyeon, T.: One-nanometer-scale size-controlled synthesis of monodisperse magnetic iron oxide nanoparticles. Angew. Chem. Int. Ed. **44**, 2872–2877 (2005)
73. Pinna, N.: The "benzyl alcohol route": An elegant approach towards organic-norganic hybrid nanomaterials. J. Mater. Chem. **17**, 2769–2774 (2007)
74. Pinna, N., Antonietti, M., Niederberger, M.: A novel nonaqueous route to V_2O_3 and Nb_2O_5 nanocrystals. Colloids Surf., A **250**, 211–213 (2004)
75. Pinna, N., Bonavita, A., Neri, G., Capone, S., Siciliano, P., Niederberger, M.: Nonaqueous synthesis of high-purity indium and tin oxide nanocrystals and their application as gas sensors. Proceedings of the IEEE Sensors 2004 pp. 192–195 (2004)

76. Pinna, N., Garnweitner, G., Beato, P., Niederberger, M., Antonietti, M.: Synthesis of yttria-based crystalline and lamellar nanostructures and their formation mechanism. Small **1**, 112–121 (2005)
77. Pinna, N., Grancharov, S., Beato, P., Bonville, P., Antonietti, M., Niederberger, M.: Magnetite nanocrystals: Nonaqueous synthesis, characterization, and solubility. Chem. Mater. **17**, 3044–3049 (2005)
78. Pinna, N., Neri, G., Antonietti, M., Niederberger, M.: Nonaqueous synthesis of nanocrystalline semiconducting metal oxides for gas sensing. Angew. Chem. Int. Ed. **43**, 4345–4349 (2004)
79. Polleux, J., Antonietti, M., Niederberger, M.: Ligand and solvent effects in the nonaqueous synthesis of highly ordered anisotropic tungsten oxide nanostructures. J. Mater. Chem. **16**, 3969–3975 (2006)
80. Polleux, J., Gurlo, A., Barsan, N., Weimar, U., Antonietti, M., Niederberger, M.: Template-free synthesis and assembly of single-crystalline tungsten oxide nanowires and their gas-sensing properties. Angew. Chem. Int. Ed. **45**, 261–265 (2006)
81. Polleux, J., Pinna, N., Antonietti, M., Niederberger, M.: Growth and assembly of crystalline tungsten oxide nanostructures assisted by bioligation. J. Am. Chem. Soc. **127**, 15,595–15,601 (2005)
82. Rao, C.N.R., Raveau, B.: Transition metal oxides. VCH Publishers Inc., New York (1995)
83. Redl, F.X., Black, C.T., Papaefthymiou, G.C., Sandstrom, R.L., Yin, M., Zeng, H., Murray, C.B., O'Brien, S.P.: Magnetic, electronic, and structural characterization of nonstoichiometric iron oxides at the nanoscale. J. Am. Chem. Soc. **126**, 14,583–14,599 (2004)
84. Ruhland, B., Becker, T., Muller, G.: Gas-kinetic interactions of nitrous oxides with SnO_2 surfaces. Sens. Actuators B **50**, 85–94 (1998)
85. Ruiz, A.M., Sakai, G., Cornet, A., Shimanoe, K., Morante, J.R., Yamazoe, N.: Cr-doped TiO_2 gas sensor for exhaust NO_2 monitoring. Sens. Actuators B **93**, 509–518 (2003)
86. Sá Ferreira, R.A., Karmaoui, M., Nobre, S.S., Carlos, L.D., Pinna, N.: Optical properties of lanthanide-doped lamellar nanohybrids. Chem. Phys. Chem. **7**, 2215–2222 (2006)
87. Sberveglieri, G.: Gas sensors. Kluver Academic Publishing (1992)
88. Schlogl, R., Abd Hamid, S.B.: Nanocatalysis: Mature science revisited or something really new? Angew. Chem., Int. Ed. **43**, 1628–1637 (2004)
89. Seiyama, T. (ed.): Chemical sensors: Current status and future outlook, *Chemical Sensor Technology*, vol. 1. Elsevier, Amsterdam (1988)
90. Seo, J.W., Chung, H., Kim, M.Y., Lee, J., Choi, I.H., Cheon, J.: Development of water-soluble single-crystalline TiO_2 nanoparticles for photocatalytic cancer-cell treatment. Small **3**, 850–853 (2007)
91. Shimizu, Y., Egashira, M.: Basic aspects and challenges of semiconductor gas sensors. Mrs Bulletin **24**, 18–24 (1999)
92. Si, R., Zhang, Y.W., You, L.P., Yan, C.H.: Rare-earth oxide nanopolyhedra, nanoplates, and nanodisks. Angew. Chem. Int. Ed. **44**, 3256–3260 (2005)
93. Si, R., Zhang, Y.W., Zhou, H.P., Sun, L.D., Yan, C.H.: Controlled-synthesis, self-assembly behavior, and surface-dependent optical properties of high-quality rare-earth oxide nanocrystals. Chem. Mater. **19**, 18–27 (2007)
94. Siciliano, P.: Preparation, characterisation and applications of thin films for gas sensors prepared by cheap chemical method. Sens. Actuators B **70**, 153–164 (2000)
95. Siddiquey, I.A., Ukaji, E., Furusawa, T., Sato, M., Suzuki, N.: The effects of organic surface treatment by methacryloxypropyltrimethoxysilane on the photostability of TiO_2. Mater. Chem. Phys. **105**, 162–168 (2007)

96. Son, S.U., Park, I.K., Park, J., Hyeon, T.: Synthesis of Cu_2O coated Cu nanoparticles and their successful applications to Ullmann-type amination coupling reactions of aryl chlorides. Chem. Commun. pp. 778–779 (2004)
97. Song, H.T., Choi, J.s., Huh, Y.M., Kim, S., Jun, Y.w., Suh, J.S., Cheon, J.: Surface modulation of magnetic nanocrystals in the development of highly efficient magnetic resonance probes for intracellular labeling. J. Am. Chem. Soc. **127**, 9992–9993 (2005)
98. Soulantica, K., Erades, L., Sauvan, M., Senocq, F., Maisonnat, A., Chaudret, B.: Synthesis of indium and indium oxide nanoparticles from indium cyclopentadienyl precursor and their application for gas sensing. Adv. Funct. Mater. **13**, 553–557 (2003)
99. Sun, F., Cai, W., Li, Y., Jia, L., Lu, F.: Direct growth of mono- and multilayer nanostructured porous films on curved surfaces and their application as gas sensors. Adv. Mater. **17**, 2872–2877 (2005)
100. Sun, S.H., Zeng, H., Robinson, D.B., Raoux, S., Rice, P.M., Wang, S.X., Li, G.X.: Monodisperse MFe_2O_4 (M = Fe, Co, Mn) nanoparticles. J. Am. Chem. Soc. **126**, 273–279 (2004)
101. Trentler, T.J., Denler, T.E., Bertone, J.F., Agrawal, A., Colvin, V.L.: Synthesis of TiO_2 nanocrystals by nonhydrolytic solution-based reactions. J. Am. Chem. Soc. **121**, 1613–1614 (1999)
102. Wang, H., Uehara, M., Nakamura, H., Miyazaki, M., Maeda, H.: Synthesis of well-dispersed Y_2O_3:Eu nanocrystals and self-assembled nanodisks using a simple non-hydrolytic route. Adv. Mater. **17**, 2506–2509 (2005)
103. Wang, Y.L., Jiang, X.C., Xia, Y.N.: A solution-phase, precursor route to polycrystalline SnO_2 nanowires that can be used for gas sensing under ambient conditions. J. Am. Chem. Soc. **125**, 16,176–16,177 (2003)
104. Williams, D.: Solid state gas sensors. Adam Hilger, Bristol (1987)
105. Willinger, M., Neri, G., Rauwel, E., Bonavita, A., Micali, G., Pinna, N.: Vanadium oxide sensing layer grown on carbon nanotubes by a new atomic layer deposition process. Nano Lett. **8**, 4201–4204 (2008)
106. Xiong, H.M., Liu, D.P., Xia, Y.Y., Chen, J.S.: Polyether-grafted ZnO nanoparticles with tunable and stable photoluminescence at room temperature. Chem. Mater. **17**, 3062–3064 (2005)
107. Xiong, H.M., Wang, Z.D., Liu, D.P., Chen, J.S., Wang, Y.G., Xia, Y.Y.: Bonding polyether onto ZnO nanoparticles: an effective method for preparing polymer nanocomposites with tunable luminescence and stable conductivity. Adv. Funct. Mater. **15**, 1751–1756 (2005)
108. Xu, C., Tamaki, J., Miura, N., Yamazoe, N.: Grain-size effects on gas sensitivity of porous SnO_2-based elements. Sens. Actuators B **3**, 147–155 (1991)
109. Yamazoe, N.: New approaches for improving semiconductor gas sensors. Sens. Actuators B **5**, 7–19 (1991)
110. Yamazoe, N.: Toward innovations of gas sensor technology. Sens. Actuators B **108**, 2–14 (2005)
111. Yamazoe, N., Miura, N. (eds.): Some basic aspects of semiconductor gas sensors, *Chemical Sensor Technology*, vol. 4. Kodansha, Tokyo (1992)
112. Yuan, Q., Ravikrishna, R., Valsaraj, K.T.: Reusable adsorbents for dilute solution. 5. Photodegradation of organic compounds on surfactant-modified titania. Sep. Sci. Technol. **24**, 309–318 (2001)
113. Yuhas, B.D., Zitoun, D.O., Pauzauskie, P.J., He, R., Yang, P.: Transition-metal doped zinc oxide nanowires. Angew. Chem. Int. Ed. **45**, 420–423 (2006)
114. Zhang, L., Djerdj, I., Cao, M., Antonietti, M., Niederberger, M.: Nonaqueous sol-gel synthesis of nanocrystalline $InNbO_4$ visible light photocatalyst. Adv. Mater. **19**, 2083–2086 (2007)

115. Zhang, Q., Chou, T., Russo, B., Jenekhe, S., Cao, G.: Aggregation of ZnO nanocrystallites for high conversion efficiency in dye-sensitized solar cells. Angew. Chem. Int. Ed. **47**, 2402–2406 (2008)
116. Zhang, Q., Chou, T.P., Russo, B., Jenekhe, S.A., Cao, G.: Polydisperse aggregates of ZnO nanocrystallites: A method for energy-conversion-efficiency enhancement in dye-sensitized solar cells. Adv. Funct. Mater. **18**, 1654–1660 (2008)
117. Zhao, F., Yuan, M., Zhang, W., Gao, S.: Monodisperse lanthanide oxysulfide nanocrystals. J. Am. Chem. Soc. **128**, 11,758–11,759 (2006)
118. Zhu, J., Yang, J., Bian, Z.F., Ren, J., Liu, Y.M., Cao, Y., Li, H.X., He, H.Y., Fan, K.N.: Nanocrystalline anatase TiO_2 photocatalysts prepared via a facile low temperature nonhydrolytic sol-gel reaction of $TiCl_4$ and benzyl alcohol. Appl. Catal., B **76**, 82–91 (2007)

Chapter 9
Summary, Conclusion and Outlook

In the last few years, the number of synthesis approaches to metal oxide nanoparticles and nanostructures reported in literature almost exploded. They gave access to a large and rapidly growing collection of oxide-based nanoparticles with a wide range of compositions, monodisperse or well-defined crystallite sizes, sophisticated crystallite shapes, and with complex assembly properties. In contrast to aqueous systems, in which smallest changes in the experimental conditions result in alteration of the products, nonaqueous procedures are very robust within one system. Therefore, most of these processes are highly reproducible, easy to scale-up to gram quantities (or more) and applicable to a broad family of metal oxides. Consequently, the nonaqueous routes summarized in this book, whether they do or do not involve the use of surfactants, offer a unique opportunity not only to chemists, but also to physicists, materials scientists, and engineers to find the appropriate synthesis method for a targeted material with the desired properties.

In comparison to aqueous sol-gel chemistry, the synthesis of metal oxide nanoparticles in organic solvents under exclusion of water provides some peculiar features, which allow for better control over particle size, shape, crystallinity, and surface properties. The main advantage of nonaqueous chemistry lies in the accessibility of highly crystalline products at moderate temperatures. This feature is strongly related to the fact that switching from aqueous sol-gel chemistry and its high reactivity of water to nonhydrolytic processes, mainly based on the reactivity of the oxygen-carbon bond, drastically decreases the reaction rates and leads to controlled crystallization. At the same time, the chemistry of the oxygen-carbon bond is well-known from organic chemistry and therefore, these routes open up the possibility to adapt reaction principles from organic chemistry to the synthesis of inorganic nanomaterials. The organic species play a manifold role during nanoparticle formation, not only determining the crystal size, shape, and assembly properties, but also the composition, the crystallization pathway, and thus the crystal structure, and, in some cases, providing a template-like matrix for the growth of dimensionally reduced nanomaterials. Whereas the morphological properties of nanomaterials are controlled by the "surfactant" properties of the organics, e.g., by their ability to cap specific crystal faces and prevent them from further growth, control of composition or crystal structure is mainly a consequence of the organics acting as oxygen source.

We explained the main chemical mechanisms so far known to lead to oxide formation under nonaqueous conditions. It is interesting to note that in spite of the large number of published methodologies, it is in principle possible to summarize the formation of most of the metal oxide nanostructures in six distinct mechanisms: i) alkyl halide elimination, ii) ether elimination, iii) condensation of carboxylate groups (ester and amide eliminations), iv) C-C coupling of benzylic alcohols and alkoxide precursors, v) aldol/ketimine condensation, and vi) oxidation of metal nanoparticles. Although these mechanisms provide fundamental answers to questions like where does the oxygen for the oxide formation come from, how the oxygen is transferred from the source to the metal center, or which organic species are formed during nanoparticle growth that potentially act as coordinating, size- and shape-controlling ligands, they do not offer any solutions concerning the relationship between a particular synthesis system and the particle morphology. As a matter of fact, the morphology of the final product strongly depends on the precursor and solvent used, i.e., metal oxides with the same composition and crystal structure, however obtained from different precursors and/or solvents, are often characterized by different particle sizes and shapes. Accordingly, the study of the organic species, their reactivity and their interaction with the growing inorganic crystals during the whole reaction pathway is a prerequisite on the way to understanding a particular reaction system. As if the study of the organic-inorganic interface at nanoscale wouldn't be complicated enough, another difficulty arises from the fact that also organic species that are not present in the initial starting solution but are formed *in situ* during the reaction course can influence the final characteristics of the sample. The consequence of the (maybe a bit) underestimated complexity of nanoparticle formation, including chemical mechanisms as well as crystallization processes, is that the ultimate goal of a rational synthesis strategy is not in sight. It is still impossible to prepare a certain compound on the nanoscale with a desired composition, structure, size and shape, or even properties, intentionally and in a predicted way.

However, research on metal oxide nanoparticles and oxide-based organic-inorganic nanocomposites not only includes synthetic work, but also the characterization of the structural, chemical and physical properties, assembly into 1-, 2- and 3-dimensional superstructures, and application in various fields of technology. Although most of the research efforts are still dedicated to the development of new synthesis routes, there is no doubt that the in-depth investigation of the physical and chemical properties represents a key step on the way to a faster implementation of inorganic nanomaterials into technological applications, and thus has to be intensified.

Figure 9.1 summarizes some of the current and future challenges in the research field of metal oxide nanoparticles and nanostructures, divided into six topics including synthesis, characterization, assembly/processing, chemical formation mechanisms, crystallization mechanisms and applications. Whereas some of these topics have already been in the focus of intense research for several years, other problems and questions just recently moved into the center of interest.

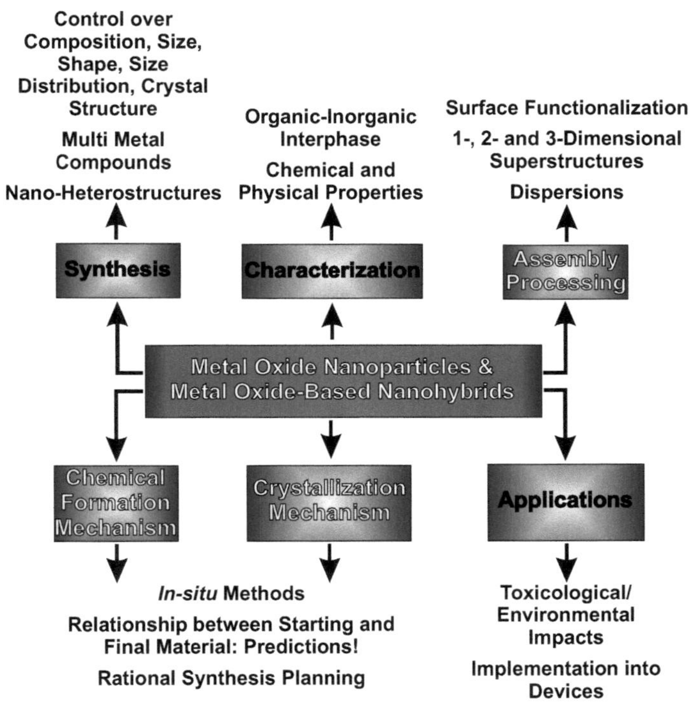

Fig. 9.1. Major challenges in the research area of metal oxide nanoparticles and oxide-based organic-inorganic nanocomposites

In the synthesis development, questions regarding control over composition, size, size distribution, shape, and crystal structure have to be continuously addressed, but the sophistication of the targeted nanoparticles has to be extended to multi metal oxide and doped metal oxide nanoparticles, nano-heterostructures containing sections of compositionally different inorganic materials and organic-inorganic nanocomposites with unprecedented combinations of organic-inorganic components with a similarly strong focus on the functionality of the organic part. It will also be important to search for "greener" synthesis procedures and to work on the industrial up-scale of the laboratory methodologies. However, there is no doubt that the primary challenge lies in the development of synthesis concepts based on a rational strategy. One has to be able to go beyond trial-and-error experiments to save time and resources.

The characterization of the as-synthesized nanomaterials becomes increasingly important, also with respect to establishing structure-property relationships. The structural characterization has to be extended beyond common laboratory techniques to e.g., synchrotron X-ray diffraction and neutron scattering (for the study of poorly crystalline nanomaterials and local structural order, determination of magnetic structures and spin dynamics). Special attention has to be directed towards the characterization of the organic-inorganic interphase (on the surface of the nanoparticles as well as within the organic-inorganic composites). In addition, there is a big backlog to study the chemical and physical properties of metal oxide nanoparticles and nanostructures obtained in organic solvents.

In the field of assembly and processing, major challenges include the surface functionalization of metal oxide nanoparticles to control their agglomeration and their assembly properties as well as to tailor their binding affinities towards other surfaces/particles and molecules, use of metal oxide nanoparticles as building blocks for the fabrication of 1-, 2-, and 3-dimensional, possibly hierarchical, architectures over several length scales, controlled patterning or deposition at specified locations, direct processing of metal oxide nanoparticles into bulk materials (thin films, ceramics, coatings, electrodes, sensors, catalysts, ...). For almost all these tasks, the first and fundamental step is the production of nanoparticle dispersions with high solid content.

To gain new insights into the crystallization and formation mechanisms of metal oxide nanoparticles it will be absolutely essential to strengthen the use of *in situ* methods under experimental conditions (e.g., neutron scattering for the study of dynamical phenomena), to determine and monitor all species in a reaction batch, initially present and *in situ* formed during the synthesis, elucidation of the chemical role of the different species in solution on a molecular level at all stages of nanoparticle growth, and correlation between a synthesis system, i.e., precursor(s) and solvent(s), and final morphological (size, shape, size distribution) and structural (crystal structure) characteristics. Furthermore, the liquid-solid interphase during all stages of crystallization has to be studied, crystallization processes have to be investigated using *in situ* techniques like neutron diffraction in real time under genuine reaction conditions or EXAFS (extended X-ray absorption fine structure) and EDXRD (energy-dispersive X-ray diffraction) investigations for the study of crystallization kinetics. An integral aspect on the way to understand nanoparticle formation will be the use of computer modelling.

Last but not least, metal oxides have a huge potential for new and emerging applications. To fully exploit these possibilities, metal oxide nanoparticles, like nanomaterials in general, have to be investigated regarding their health, safety, and environmental aspects (including potential long-term effects). Other challenges include the combination of nanoparticle technology with current industrial top-down approaches, implementation of nanoparticles into (nano)devices, finally leading to one-particle devices, combination of metal oxide nanoparticles with other classes of nano- and bulk materials (composites, hybrids, ...), and study of the stability of nanostructured materials during operation, e.g., thermal, chemical, and structural stability on the long term. For large scale applications, scalability, reproducibility and cost-effective production have to be improved.

As a summary of this long list, one can say that in spite of all the progress achieved in the fascinating field of nanoparticle research on metal oxides, we are still at the very beginning ...

Index

Biographies

Markus Niederberger studied chemistry at the Swiss Federal Institute of Technology (ETH) Zürich, where he also received his Ph.D. in 2000. After a postdoctoral stay at the University of California at Santa Barbara, he became group leader at the Max Planck Institute of Colloids and Interfaces at Potsdam in 2002. Since 2007 he is Assistant Professor in the Department of Materials at the ETH Zürich. His primary research interests are the development of general synthesis concepts for inorganic nanoparticles, investigation of formation and crystallization mechanisms, nanoparticle assembly, and nanoparticle-polymer composites. He is also Editorinchief of Nanoscale.

Nicola Pinna studied physical chemistry at the Université Pierre et Marie Curie (Paris). He received his Ph.D. in 2001, and in 2002, he moved to the Fritz Haber Institute of the Max Planck Society (Berlin). In 2003, he joined the Max Planck Institute of Colloids and Interfaces (Potsdam). In 2005, he moved to the Martin Luther University, Halle-Wittenberg, as an Assistant Professor of Inorganic Chemistry. Since 2006 he is senior researcher in the Department of Chemistry and CICECO of the University of Aveiro (Portugal) and since 2009 he is also Assistant Professor at the Seoul National University. His research activity is focused on the synthesis of nanomaterials by nonaqueous sol-gel routes, their characterization, and the study of their physical properties.

MIX
Papier aus verantwortungsvollen Quellen
Paper from responsible sources
FSC® C105338

If you have any concerns about our products,
you can contact us on
ProductSafety@springernature.com

In case Publisher is established outside the EU,
the EU authorized representative is:
Springer Nature Customer Service Center GmbH
Europaplatz 3, 69115 Heidelberg, Germany

Printed by Libri Plureos GmbH
in Hamburg, Germany